DOMESTIC
GEESE AND DUCKS

*A Complete and Authentic Handbook and Guide
for Breeders, Growers and Admirers
of Domestic Geese and Ducks*

By
PAUL IVES

President: American Waterfowl Association
Editor: Cackle and Crow, the Poultrypaper, New Haven, Conn.

ILLUSTRATED BY
FRANKLANE L. SEWELL
AND
ARTHUR O. SCHILLING
AND OTHERS

1947

British Library Cataloguing-in-Publication Data
A catalogue record for this book is available from the
British Library

Poultry Farming

Poultry farming is the raising of domesticated birds such as chickens, turkeys, ducks, and geese, for the purpose of farming meat or eggs for food. Poultry are farmed in great numbers with chickens being the most numerous. More than 50 billion chickens are raised annually as a source of food, for both their meat and their eggs. Chickens raised for eggs are usually called 'layers' while chickens raised for meat are often called 'broilers'. In total, the UK alone consumes over 29 million eggs per day

According to the Worldwatch Institute, 74% of the world's poultry meat, and 68% of eggs are produced in ways that are described as 'intensive'. One alternative to intensive poultry farming is free-range farming using much lower stocking densities. This type of farming allows chickens to roam freely for a period of the day, although they are usually confined in sheds at night to protect them from predators or kept indoors if the weather is particularly bad. In the UK, the Department for Environment, Food and Rural Affairs (Defra) states that a free-range chicken must have day-time access to open-air runs during at least half of its life. Thankfully, free-range farming of egg-laying hens is increasing its share of the market. Defra figures indicate that 45% of eggs produced in the UK throughout 2010 were free-range, 5% were produced in barn systems and 50% from cages. This compares with 41% being free-range in 2009.

Despite this increase, unfortunately most birds are still reared and bred in 'intensive' conditions. Commercial hens usually begin laying eggs at 16–20 weeks of age, although production gradually declines soon after from approximately 25 weeks of age. This means that in many countries, by approximately 72 weeks of age, flocks are considered economically unviable and are slaughtered after approximately 12 months of egg production. This is despite the fact that chickens will naturally live for 6 or more years. In some countries, hens are 'force molted' to re-invigorate egg-laying. This practice is performed on a large commercial scale by artificially provoking a complete flock of hens to molt simultaneously. This is usually achieved by withdrawal of feed for 7-14 days which has the effect of allowing the hen's reproductive tracts to regress and rejuvenate. After a molt, the hen's production rate usually peaks slightly below the previous peak rate and egg quality is improved. In the UK, the Department for Environment, Food and Rural Affairs states 'In no circumstances may birds be induced to moult by withholding feed and water.' Sadly, this is not the case in all countries however.

Other practices in chicken farming include 'beak trimming', this involves cutting the hen's beak when they are born, to reduce the damaging effects of aggression, feather pecking and cannibalism. Scientific studies have shown that such practices are likely to cause both acute and chronic pain though, as the beak is a complex, functional organ with an extensive nervous supply. Behavioural evidence of pain after beak trimming in layer hen chicks has been based on the observed reduction in pecking behaviour, reduced activity and social behaviour, and increased sleep duration. Modern egg laying breeds also frequently suffer from osteoporosis which results in the chicken's skeletal system being weakened. During egg production, large amounts of calcium are transferred from bones to create egg-shell. Although dietary calcium levels are adequate, absorption of dietary calcium is not always sufficient, given the intensity of production, to fully replenish bone calcium. This can lead to increases in bone breakages, particularly when the hens are being removed from cages at the end of laying.

The majority of hens in many countries are reared in battery cages, although the European Union Council Directive 1999/74/EC has banned the conventional battery cage in EU states from January 2012. These are small cages, usually made of metal in modern systems, housing 3 to 8 hens. The walls are made of either solid metal or mesh, and the floor is sloped wire mesh to allow the faeces to drop through and eggs to roll onto an egg-collecting conveyor belt. Water is usually provided by overhead nipple systems, and food in a trough along the front of the cage replenished at regular intervals by a mechanical chain. The cages are arranged in long rows as multiple tiers, often with cages back-to-back (hence the term 'battery cage'). Within a single shed, there may be several floors contain battery cages meaning that a single shed may contain many tens of thousands of hens. In response to tightened legislation, development of prototype commercial furnished cage systems began in the 1980s. Furnished cages, sometimes called 'enriched' or 'modified' cages, are cages for egg laying hens which have been designed to overcome some of the welfare concerns of battery cages whilst retaining their economic and husbandry advantages, and also provide some of the welfare advantages of non-cage systems.

Many design features of furnished cages have been incorporated because research in animal welfare science has shown them to be of benefit to the hens. In the UK, the Defra 'Code for the Welfare of Laying Hens' states furnished cages should provide at least 750 cm^2 of cage area per hen, 600 cm^2 of which should be usable; the height of the cage other than that above the usable area should be at least 20 cm at every point and no cage should have a total area that is less than 2000 cm^2. In addition, furnished cages should provide a nest, litter such that pecking and scratching are possible, appropriate perches allowing at least 15 cm per hen, a claw-shortening device, and a feed trough which may be used without restriction providing 12 cm per hen. The practice of chicken farming continues to be a much debated area, and it is hoped that in this increasingly globalised and environmentally aware age, the inhumane side of chicken farming will cease. There are many thousands of chicken farms (and individual keepers) that treat their chickens with the requisite care and attention, and thankfully, these numbers are increasing.

From an oil painting by Franklane L. Sewell

May the gift of heavenly peace
And glory for all time

Keep the boy Tom, who, tending geese,
First made the nursery rhyme.
—*A Ballad of Nursery Rhyme,
Stanza 6, Robert Graves.*

FOREWORD

Many hundreds of books and literally millions of words have been written on chickens from the time Cato, the old Roman writer on agriculture 100 years B.C., told the Roman farmers how to select and manage poultry on the farm, until the present day.

Volume after volume, special magazines and untold articles have been written and published on the turkey and its breeding, care and management. Ducks have received plenty of attention from the poultry authorities and educators and squab raising has received the attention of many more.

But the Goose, as old as history and as authentic, consistently dependable and potentially profitable as any live thing on the farm, has pretty much paddled its own canoe with little if any publicity, propaganda or praise from the writers and educational agencies, still survives and pursues the even and unspectacular tenor of its way; which in itself demonstrates its value as a creature well able to maintain itself and render its service without too much dependence on man.

Geese are intriguing creatures; wise and canny to an almost unbelievable degree except to those who know them. There is nothing more inappropriate than the saying "silly as a goose"; no domestic fowl knows as much. There is no fowl that requires as little care, housing or attention and none that has the possibilities for as much profit on as little investment of time, labor and cash.

This book then, is designed to give in simple words and briefly, something of the possibilities of pleasure, satisfaction and profit in goose raising; as an inexpensive and fascinating hobby, an interesting and attractive addition to the activities of the estate or country home, as a paying sideline on the general farm or as a profitable business enterprise.

CONTENTS

PART ONE—GEESE

PART ONE
GEESE

CHAPTER I

History, Habits, Tradition and General Possibilities in Geese

Back to the beginning of history and back of that into folklore and mythology, runs the story of the goose.

All through a lifetime in which poultry has been the major interest in my mind, geese have intrigued my attention and tightly gripped my fancy. There is no bird in all the domestic list that compares in intelligence, dignity, self-assurance and the ability to take care of itself, with the goose.

Geese can be raised at less expense, with less equipment, less boughten feed, less labor and less attention than any other feathered creature on the farm and since the beginning of organized agriculture have geese been a recognized part of the farmstead livestock.

In the primitive days in New England when the husbandman supported himself from his own farm, they furnished the feathers for the beds and down comfortables which kept him and his family snug and comfortable in zero-cold bedrooms through the winter nights—and even aided bashful courters back in eves of bundling. They furnished the pens with which Benj. Franklin learned to write. They provided the customary roast goose for Christmas dinner and the jars of goose oil which softened up the bronchial tubes of "croupy" children when the doctor was snow-bound twenty miles away; and furnished the "shortening" for the traditional mince and dried apple pies of New England farm breakfasts.

The goose was a useful bird and in those days was held in far greater esteem than the chicken that took much and more expensive feed and furnished nothing but a few eggs in May and June and an occasional Sunday dinner. His food was nothing but grass 'till winter froze the streams and killed the verdure and then a handful or two of corn a day and plenty of hay or "cornfodder" was all he needed to live comfortably 'till green grass grew again.

He needed no shelter and scorned it if furnished. His mate laid in April and the goslings came in May, nicely timed for the first young and tender grass which was all the food required. When Fall came and the nights grew cold, home grown corn twice a day put on the rich flesh that made him prime for food. Of course the goose was a valuable bird on all New England farms. He returned more for what he received than any other live thing the farm produced; and our early New England ancestors were a careful, close and canny lot; so naturally the goose was valued and kept on nearly every farm.

But mass production and specialization were both on the way in; and general or subsistance farming on the way out. Industry began to grow and concentrate in the cities. Farmers started to raise the food to feed the workers in industry and buy the goods that industry turned out. The dairy cow crowded other livestock off some farms; chickens began to find a ready sale in the towns and cities and chickens were improved by breeding, care and selection and eggs became a "cash crop."

Some farmers specialized on fruit, some on poultry and eggs, some vegetables, some wool and mutton and some on dairy products and the special interest on each farm gradually crowded lesser interests out.

Each farmer became a specialist to a larger or lesser

degree; he "put all his eggs in one basket" and is only now beginning to realize that producing just one thing and selling it at wholesale and buying everything he needs at retail, gets him nowhere slowly but surely as the cost of transportation and distribution has, as time has passed, increased to thirty to fifty per cent of the price the consumer pays.

So now the trend is back to a more general type of farming and the agricultural teaching is leading the trend; but in the shuffle, our faithful friend, the goose, lost out. She couldn't bring a steady income the year round and farmers had to have the "quick dollar" to meet their weekly bills.

But, did you ever notice that the idea which is sound and right never, never dies? That the farm practice that is founded on economics always, in the end, prevails? That the bird, animal, tree or plant that is fundamentally in accord with its environment never becomes extinct?

Back in the days of Cato and Varro 2000 years ago, the Roman farmers were told "practice fallow and rotation"—"Seeds should be carefully selected"—"Certain plants affect certain soils"—"Study the points of cattle"—"Recruit your herd with fresh blood"—"Compensate the land by planting legumes"—*"Sow for the geese,* endive which keeps green wherever there is water." There is really nothing new in principles of farming. Will any county agent advise against any of the above? Methods change—but sound principles— never!

The Rev. Edmund Saul Dixon writing in 1851, speaks highly of the goose as follows: ". . . The value and usefulness of Geese is scarcely calculable. We omit what is owing to them, as having furnished the most powerful instrument wielded by the hand of Man (the pen). But in a mere material point of view, and reckoning on the very smallest scale, we will suppose that a village green supports only fifty brood Geese. The owners of these would be dissatisfied if they got but ten young ones from each in the year, besides Eggs; this gives five hundred Geese per annum, without taking the chance of a second brood. Multiply five hundred by the number of village greens in the kingdom, and we still form a very inadequate estimate of the impor-

Business Toulouse on the farm of John Reid, Massachusetts.

tance of the bird. And all this with scarcely any outlay. The little trouble they demand, of being secured at night and let out in the morning, of setting the Geese, and 'pegging' the Goslings, is a source of amusement and interest to thousands of aged and infirm persons, in whose affections their Geese stand second only to their children and relations. What a pity it is that such cheerful commons should be ever converted into barren thickets and damp Pheasant covers, to afford a school for young sportsmen and rural policemen to practice their several arts in.

"The only damage they do, lies in the quantity of

food they consume; the only care they require, is to be
saved from thieves and starvation. All the fears and
anxieties requisite to educate the Turkey and prepare
it for making a proper appearance at table, are with
them unnecessary; grass by day, a dry bed at night,
and a tolerably attentive mother, being all that is required. Roast goose, fatted, of course, to the point of
repletion, is almost the only luxury that is not thought
an extravagance in an economical farm-house; for
there are the feathers, to swell the mistress' accumulating stock of beds; there are the drippings to enrich the
dumpling, pudding, or whatever other farinaceous food
may be the fashion of the country for the servants to
eat; there are the giblets, to go to market or make a
pie for a special occasion, and there is the wholesome,
solid, savoury flesh for all parties in their due proportion.

"They are accused, by some, of rendering the spots
where they feed offensive to other stock; but the secret
of this is very simple. A Horse bites closer than an Ox,
a Sheep goes nearer to the ground than a Horse, but,
after the sharpest shaving by Sheep, the Goose will
polish up the turf, and grow fat upon the remnants of
others. Consequently, where Geese are kept in great
numbers on a small area, little will be left to maintain
any other grass-eating creature. But if the commons
are not short, it will not be found that other grazing
animals object to feed either together with, or immediately after a flock of Geese."

Charles O. Flagg at that time director of the Rhode
Island Agricultural Experiment Station who must
have been himself a lover of geese, wrote the following
regarding them in his report to the General Assembly
for 1897. It is so beautifully written and indicates such
a thorough knowledge of the subject that I cannot re-
frain from presenting it to what I hope may be a much

wider audience than when originally written. I wish to express appreciation and thanks to my friend Dr. H. O. Stuart, head of the Poultry Department of the Rhode Island State Agricultural College for making this old report, now out of print and very rare, available to me.

"At what time in the world's history the goose became a domestic animal literature fails to inform us: but, that together with the waymarks of sculpture, art and science, which indicate the progressive march of humanity through the centuries past, shows us that, in common with the barn-yard fowl, the goose has been a servant of man from the earliest times. Ancient literature ought, rightly, to tell us more about our subject than it does, for, since sometime in the fourth century, the goose has provided the scribes with quills wherewith to record, for our instruction, the great and small events of history; the noble and the base in the manhood of all times since then; the rise and fall of kings and empires; the constant struggle of truth with error, and to picture for us the customs and manners, the loves and sorrows, and the faults and foibles of our ancestors. Though a feather seems a 'trifle, light as air,' yet the feathers of the goose have had much to do with the physical and mental comfort of mankind, even far back in the centuries, and although no marble monument has been raised to do her honor, yet the service she has rendered will live so long as written language shall endure.

"If we inquire as to the origin of our breeds of geese, we find that there are one or two writers who contend that the wild prototype of the domestic goose does not now exist, and cite the camel as an analogous case. They give as a reason therefore the fact that the domestic goose is the only bird of its tribe systematically polygamous—all the known wild varieties mating in pairs for breeding, and this applies even to the wild

Canada goose now in domestication. The large majority of authorities, however, is united in the belief that
the common domestic goose is descended from the in-
digenous wild goose of the British Islands known as the
greylag goose. The name is sometimes given as 'grey-
leg' or 'grey-legged goose,' but lacks point, as the bird
is grey in plumage, while the legs are yellowish in
color; at the same time the term *lag* had no reasonable
explanation until in 1870, Prof. Skeat suggested that
the appellation was given because this goose lagged behind when other varieties of wild geese migrated, which
was the case in early times. This goose (*Anser ferus*)
is found throughout middle and northern Europe and
Asia, migrating to marshes and low grounds or islands
in the north during the summer season. The greylag
goose breeds in more southern latitudes than the
Canada goose, and seldom, if ever, visits the far north
as does the latter. The size is a little larger than that
of the Canada goose (*Anser Canadensis*), in a wild
state specimens often weighing ten pounds. While
many of the goose family, which includes some forty
species, live largely upon insect, animal and vegetable
life, found in or close by the bodies of water which they
frequent, the domestic goose is eminently a grazing
animal. In early life its most rapid growth is made
upon a pasture of short nutritious grasses, supple-
mented with a little grain. Good pasturage, with water
for drinking only, will produce well grown geese as
easily as sheep or cattle.

"Darwin says: 'hardly any other anciently domesticated bird or quadruped has varied so little as the
goose,' and Hewitt says: 'My opinion is that the greylag is probably the original stock from which all, or at
least most of our common varieties sprang, my idea
being based on the fact that frequently we see the most
unquestionable tendency to "breed back," a bird hav-

ing all (or nearly all) of the traits of character of the greylag, even when the parent birds did not exhibit the slightest resemblance.'

"The greylag goose is thus described: 'The bill is pale flesh-colored yellow, somewhat lighter than the legs; the nail, at its extremity, being white. The neck and back are ash grey; wings, a light brown, the edges of feathers running into a lighter tint, while the lesser wing-coverts are of bluish grey, in contrast to the darker hue of the wing generally, a peculiarity that often serves to distinguish this species in both adult and immature specimens; the breast and front of the neck light ash grey, the former being lightly barred with transverse markings, tail coverts and under part of the body, white, tail feathers a dull brown with a white margin. The 'curl' of the neck feathers, so remarkable a feature in the domestic goose, is strongly marked in this species.'

"The characteristic last named, the curled or twilled appearance of the neck feathers, is a very prominent marking in both the Embden and Toulouse geese of today. It is, however, almost wholly absent in the case of the White China and Brown China geese, in which the tendency toward an erect carriage is marked, and, taken together with the prominent 'knob' or protuberance at the base of the upper mandible, the harsher, shriller voice and greater prolificacy, points to a different origin or to much greater changes in characteristics, through domestication and selection, than in the case of the Embden and Toulouse breeds. Naturalists give the Chinese goose the name of *Anser cygnoides* and Wright quotes Blyth as authority that the common domestic goose of India is a hybrid between this goose and the greylag, and says: 'It is very remarkable that these hybrids appear perfectly prolific and perpetuate the cross with a little care; not, as is usual with crosses, reverting to either of the parent races.' Is not this hybrid the original foundation of the African goose as bred at the present time? The names African and India are used interchangeably by many goose breeders, and the characteristics of the breed show modified Brown China traits. The general color of the plumage is much the same, and the dark brown stripe on the back of the head and neck, which is so strongly characteristic of the China, is retained, and also the knob at the base of the upper mandible, although considerably reduced in size in proportion to the relative size of the birds. The dewlap, or pendant skin under the throat, which Wright and Brown say is a distinguishing characteristic of the Chinese goose, is now required fully developed in African geese, and *is not now* countenanced in the pure Chinese breeds.

"Darwin says: 'Although the domestic goose certainly differs somewhat from any known wild species, yet the amount of variation which it has undergone, as compared with that of most domestic animals, is singularly small. This fact can be partially accounted for by selection not having come largely into play. Birds of all kinds, which present many distinct races, are valued as pets or ornaments; no one makes a pet of the goose, the name, indeed, in more languages than one, is a term of reproach. The goose is valued for its size and flavor, for the whiteness of its feathers, which adds to their value, and for its prolificness and tameness. In all these points the goose differs from the wild parent form, and these are the points which have been selected. Even in ancient times the Roman gourmands valued the liver of the *white* goose; and Pierre Belon, in 1555, speaks of two varieties, one of which was larger, more fecund, and of a better color than the other; and he expressly states that good managers at-

tended to the color of their goslings, so that they might
know which to preserve and select for breeding.'

"White is evidently a color developed by domestication and selection. The estimation in which white
birds were held by the Romans no doubt led to their
preservation as breeding stock, but the custom of
plucking live geese for the feathers, followed probably
for hundreds of years, has, no doubt, had its influence,

A fine pair of Brown Chinese and their five young. Owned by Har-
riman A. Reardon, Hudson, Mass. (Note one pure white sport.)

as bird and poultry keepers know that a white feather is often produced in place of a colored one pulled out. It has been said of the common domestic geese of England, that 'the ganders are usually white, or with a preponderance of that color, while the geese have various shades of ash grey and a dull leaden brown mixed with it; a preference is often expressed for those that have no white whatever, excepting only on the lower part of the body.' Wright refers to this preference of color in mating, when for certain reasons he advises the crossing of a Toulouse gander with Embden geese he says: 'It also affords some amusement to the owner, as it altogether upsets at once the theory of many old farm mistresses, that the gander is the white bird and the geese particolored.' The breeding of white geese has also had some encouragement because of the greater value of white feathers as compared with colored or feathers of mixed color, and because the dressed bird has a brighter and cleaner appearance, more pleasing to the eye than that of a dark feathered bird, and which therefore helps its sale in the market. An English authority says: 'All white aquatic poultry are considered to dress i.e. to "pluck" of a clearer and better appearance than the particolored or dark feathered birds, more especially whilst young. This arises from the patches, where the dark feathers grew, showing even after being carefully plucked, more particularly if the plumage at the time they are killed happens to be immature. Although when roasted no difference is perceptible, yet a clear skinned bird always commands the most ready sale.' This partiality of the public for that which presents a *fine appearance* is manifest in the development of the white breeds of fowls so popular with those engaged in the raising of broilers; the Pekin duck in this country, and the Aylesbury duck in England.

"Aside from color, domestication and selection have

changed the goose in respect to size and fecundity. From the wild type, weighing at maturity about ten pounds each, have been developed, in the course of time, birds weighing on exhibition sixty pounds per pair, and thirty-eight pounds for a single male bird. These green geese at twelve weeks old to weigh from twelve to fifteen pounds each, and at four or five months old to reach eighteen to twenty or more pounds, dressed weight.

"As before stated, the domestic goose, of all the goose family, is the only one where the gander quite regularly mates with more than one goose. He, however, seldom mates with more than four geese, and often with less, usually having one favorite whom he guards more jealously than the others, and whose nest he is ever ready to defend against all comers. The wild greylag goose lays generally from five to eight eggs, and has been known to lay twelve to fourteen, while some varieties of the domestic goose, if not allowed to sit, will sometimes lay sixty or more eggs in a single season. Selection of breeding stock and feeding have much to do with the egg production. Rankin says in regard to the laying qualities of African geese: 'Thirty years ago I rarely had a bird that would lay over thirty eggs; now they often lay sixty, and occasionally more.' A California correspondent to a poultry paper, states that he kept one pair of Toulouse geese, and in 1885 the goose laid 65 eggs, of which number 53 were set under hens and every egg hatched. As a rule the Chinese geese lay more eggs than other varieties.

"The wild greylag goose interbreeds with the domestic goose, and the progeny is fertile.

"The wild Canada goose is quite readily domesticated, and the ganders will usually mate the second or third year with a domestic goose. A dark colored female, usually Toulouse or African, is selected for such mat-

ing, and the progeny is the 'mongrel' goose so highly
prized for the table, and which always far exceeds the
price of other geese in the market. The wild female is
seldom mated with the domestic gander, as she lays but
few eggs, and the production of 'mongrels' from such
matings is very limited and hardly profitable. The
'mongrel' progeny of either mating is sterile. Audubon
says: 'The greatest number of eggs I have found in the
nest of this species (Canada goose) was nine, which I
think is more by three than these birds usually lay in a
wild state. In the nests of those which I have had in
domesticated state I have sometimes counted as many
as eleven. Several of them, however, usually proved
unproductive. They never have more than one brood
in a season unless their eggs are removed or broken at
an early period.'

"The successful breeding and rearing of wild (Canada) geese and 'mongrels,' or hybrids between the wild
and African or Toulouse goose, is the perfection of art
in goose raising, and only those who are thoroughly
familiar with the habits and peculiarities of the wild as
well as the domestic goose, and so situated as to provide each pair of them with abundant space, including
a natural supply of water in some secluded locality in
which they may reign supreme, can hope for good success.

"The few men who are successful breeders of mongrel
geese have as a rule grown up in the business from boyhood, and have a lifelong apprenticeship combined
with infinite patience and tact.

"It has been said that no class of poultry can be produced with so little expense for shelter, food, labor, and
care, as geese. This statement is true when their habits
and requirements are thoroughly understood and met,
and it is equally true that no class of poultry can be
more disastrously unsatisfactory under opposite condi-

tions. The habits and peculiarities of the five common varieties of domestic geese bred in this country are generally uniform and resemble those of the wild goose; modified as would be expected by centuries of domestication. Whether in the course of time the Canadian gander will become polygamous, and the goose develop an egg producing capacity two to five times as great as

A flock of Canadian Geese. Courtesy Government Experimental Farm, Ottawa, Canada.

at present, is problematical, but one would say quite possible in view of the changes which selection and domestication have produced in the case of our breeds of domestic geese. In order that the novice may have some idea of the conditions to be met in successful goose rearing, before enumerating the domestic varieties we will endeavor to point out some of the peculiar habits of geese.

"We very often hear the word goose used to designate a person as silly, or to characterize some foolish action. This use of the word, as indicating a popular opinion regarding the stupidity of the bird, is resented by those familiar with their habits as owners and breeders of geese, and even some authors declare the opinion erroneous.

"A writer in the *Cornhill Magazine* says, in regard to the popular use of the word, 'It being only ignorance of the darkest hue that ventures to portray the goose as deficient in sagacity and intelligence.' Probably this erroneous popular opinion may be attributed to the one quality of timidity, and the liability of the goose to act very foolishly when frightened, more than to any other trait.

"Those who adhere to the popular conception regarding geese may not appreciate this trait of timidity, in view of the traditions one often hears regarding certain courageous and pugnacious ganders which at times have inflicted serious injuries upon strong men. Geese have thereby obtained credit for a degree of courage and a spirit which is not usually theirs, or manifest only during the breeding season and while the geese are sitting. Perhaps no other domestic fowl requires to be more quietly and carefully cared for than the goose. Undue excitement, or disturbance by visitors, strange dogs or animals, often has a very injurious effect upon them, especially in the laying season. One breeder says that he has known geese to be so badly frightened from the throwing of a few cabbages into the yard as to affect the egg production. A very nervous or fractious person does not usually have very good success in the handling of geese. They require the kindest of treatment, and the breeder should be thoroughly familiar with the individuals of his flock, and on the most intimate terms with them in order to attain the best success. With this trait of timidity is its counterpart—extreme watchfulness. Geese are ever on the alert, and one breeder asserts that geese are better than any watch-dog for giving notice of the approach of strangers during either daytime or night. One goose breeder who claimed membership in one or two societies, and occasionally came home quite late in the evening, asserted that he could always get into the house without disturbing his watch-dog, but he never in his life succeeded without arousing the gander, which gave due notice of his approach. It was this trait of watchfulness which gave to geese the credit of saving Rome from surprise and capture, through a silent and stealthy night attack of the enemy, as early as 388 B.C. Then geese were kept as sacred to the queen of the Roman goddess, Juno, which sacred-ness implied great antiquity.

"The greylag goose, in a wild state, feeds in flocks of greater or less size and always with sentinels on guard ready to sound an alarm upon the slightest approach of danger. Bishop Stanley says 'no animal biped or quadruped is so difficult to deceive or approach.' The Canada goose is equally watchful and wary of anything which threatens the safety of the flock. Audubon says 'in keenness of sight and acuteness of hearing they are perhaps surpassed by no other bird.' Their ability to distinguish between sounds made by wild animals, as the breaking of a twig by a deer or the splash of water by a turtle, and similar sounds produced by the ap-proach of the hunter, is phenomenal. When one wishes to define an undertaking as exceedingly doubtful as to profitable results, he can use a no more expressive term than to characterize it as a 'wild goose chase.' That proverbial saying has its foundation in the watchful-ness, acuteness, and capacity to look out for its own safety and that of its family possessed by the goose.

"As intimated, ganders, during the breeding season, and even the geese when sitting, or in defense of their young, manifest considerable courage and often punish intruders severely. When interfered with, they seize the intruder with the bill, strike with the wings, and sometimes scratch with the claws. They have sufficient power in the jaws to bite quite hard, and a large, full grown gander has been known to strike hard enough with the wings to break a person's arm. It is very rarely, however, that a gander kindly cared for and treated well, becomes habitually ugly so as to attack people without provocation. The ganders fight among themselves whenever one colony intrudes upon the territory of another, and their battles are severely fought, usually with the wings, one gander seizing the other by the first joint of the wing with the bill, and beating him with his wings while thus held. Unless separated at such times, they are liable to receive injury; however, where large parties run together, accustomed to each other's society, they usually understand their position and relations, so that very little, if any, difficulty is experienced from fighting.

"As has been before stated, geese are grazing animals to a greater extent than any other class of poultry. In fact they live and thrive on good pasturage and water, although of course they do not make the rapid growth that may be secured when some grain is fed; on the other hand, however, it is not possible, probably, to raise goslings on an exclusive grain diet without a liberal supply of clover, cabbage, roots, apples or some succulent vegetable food. Young goslings make the most rapid growth upon short nutritious grass and cracked corn or wheat. In a wild state, geese devour large quantities of roots of grasses and aquatic plants, which they dig from the banks and borders of streams and wash free from earth in the shallow water. Domestic geese confine themselves less to water and aquatic plants and generally feed upon pastures, preferring moist, rich localities where the grass is kept short and sweet by constant feeding and rapid growth. Tall woody grasses, which have become tough, are not relished by them. This natural habit of geese makes considerable space necessary for their successful keeping, or requires that they be provided with succulent green crops, such as rape, cabbage, sorghum, corn, oats, etc.

Three young Toulouse at three weeks. Courtesy Andrew Taylor, Government Experimental Farm, Ottawa, Canada.

"Broods of goslings of different ages, hatched and reared on the same farm, must, of necessity, be penned while young, each brood by itself, and as they go out to feed on the pasture or field each flock invariably keeps by itself. Any intruder or visitor from another flock is very unwelcome, and is scolded, bitten, and driven out of the flock by common consent. This clannish rule is peculiar to geese and very strictly enforced. Saunders says: 'If we traverse a pasture or common on which geese are kept, we find the flocks of the respective owners keeping together; and if by chance they mingle on the pond or sheet of water, they separate towards evening and retire, each flock to its own domicile. On extensive commons, where many thousands of geese are kept, the rule is scarcely ever broken; the flocks of young geese, brought up together as their parents were before them, form a united band, and thus distinct groups herd together, bound by the ties of habit.'

"An old adage, more expressive than elegant, says: 'The goose eats everything before it and poisons everything behind it,' but doubtless it had its origin in the mind of some enemy, as when geese have sufficient pasture it is not true.

"Unlike gallinaceous fowls, the goose has practically no crop, although an enlargement of the end of the gullet next the gizzard in some measure serves to hold food, consequently it feeds at very frequent intervals, and during warm weather often eats more at night than during the day-time, a point which should be remembered in feeding and caring for them.

"Geese have great constancy, another trait which is not appreciated except by those who have had considerable experience in raising them. This term applies to their attachment for each other, and also to their home and surroundings.

"The wild Canada gander usually mates with but one goose, and, once mated, is constant in his attachment to the goose of his choice so long as she is allowed to remain with him. The domestic goose seldom mates with more than three females, occasionally with less, and is almost equally constant in his adherence to the mates he has selected. If, for any reason, he is separated from his mates and placed with others, he will seldom accept them so long as his old mates are anywhere within hearing distance, and, even when they are entirely removed from the premises, it frequently takes some time before he will become reconciled to his new mates. The wild gander almost never mates the first year, and frequently not until the third season, and is much more particular about accepting a new mate if deprived of one to which he has become already attached. Young geese are not fully mature at twelve months old, and the experienced breeder never expects the best results in egg production and fertility until the second or third season.

"These peculiarities of geese are not appreciated by the novice, and because eggs fail to hatch, and poor results are attained the first season, the business of goose raising is given up when really a proper trial under suitable conditions has not been made.

"Geese become attached to the locality in which they are kept and are much disturbed when removed to a new location; hence, when such removal is necessary, or when a beginning is to be made in the keeping of geese, breeding birds should be placed in their new quarters some weeks before the laying season begins, or a good number of fertile eggs will probably not be obtained.

"Geese have a long tenure of life, far exceeding any other domestic fowl in this respect. In former times it was not uncommon for the farmer's daughter, on her wedding day, to receive, among other gifts, a goose from the old homestead, to become her property and accompany her to her new home. In some instances such geese were kept for many years, perhaps far beyond the life of the young lady to whom it was presented.

"Such a goose was exhibited at the New Jersey State Fair, in 1859 and her history, on a placard posted on the coop, read as follows: 'Madam Goose is now owned by Robert Schomp, of Reading, Hunterdon County, N. J. She has been in his possession twenty-five years, and was given to him by his grandfather, Major H. G. Schomp. Robert's father is now in his eighty-fifth year, and this goose was a gift to his mother as a part of her marriage outfit. The mate of Madam Goose was killed in the revolutionary war, being rode over by a troop of cavalry . . . In the spring of

1857 she laid six eggs, three of which were hatched, and the goslings raised. In 1858 she made seven nests and laid but two eggs, evidence perhaps of failing faculties. Her eyes are becoming dim, one having almost entirely failed. The year of her birth cannot be known, but she remains a representation of the oldentime.'

Some '44 Toulouse, January 27, 1945. From the famous Oscar Grow Strain. Courtesy Reo Mowrer, Unionville, Mo.

"William Rankin, a veteran goose breeder, cites the instance of a goose owned in Boxford, Mass., where it was the property of one family for 101 years, and was then killed by the kick of a horse. She had laid 15 eggs and was sitting on them when a stray horse approached too near the nest; she rushed off, in defense of her eggs, seized the horse by the tail, and was killed by a kick from the animal.

"The same gentleman, about 25 years ago, purchased in this State a wild gander which had been owned by one family some 50 years. A member of the family had wounded the gander by firing into a flock of wild geese, breaking his wing. The gander recovered from his injury and was kept for that number of years, without, however, mating with other geese. He is now kept and used as a decoy bird during the gunning season, and highly valued by his owner, although at least 75 years old.

"Willoughby records the instance of a goose that had reached the age of 80 years, and was at last killed for its mischievousness.

"Some goose raisers say that geese seldom get too old to be good breeders, while occasionally one prefers geese from two to five years old. Barring accidents, good geese may be profitably kept until 25 or more years old; ganders of the domestic varieties, however, are less useful after 7 or 8 years, and should be replaced with young birds. While the young gander often mates with three or four females, he usually has one particular favorite among the number, whose nest he guards more jealously than those of his other mates, and after some years he is liable to grow so inattentive to all but the favorite that many of the eggs produced prove to be infertile, and it is more economical to replace him with a younger bird. The Canada gander is, however, a pretty sure and valuable breeder for many years.

"Ganders occasionally take very peculiar freaks, such as conceiving a violent attachment for some inanimate object, as a door, stone, a cart wheel, a plow, or something of a similar nature, when they will spend the greater part of their time sitting beside it or in its company. Morris relates a number of instances where ganders have become the inseparable companions of their masters, following them about the fields, on hunting expeditions, and into the streets of a town, like the most devoted dog. He also narrates how faithfully a gander discharged the self-imposed duty of guardian and guide to an old blind woman. Whenever she went to church he directed her footsteps into safe paths by taking hold of her gown with his bill, and during the service he nipped the grass in the cemetery close by until she required his services as guide to return home.

"Geese are peculiar, in that both sexes are feathered exactly alike. Consequently there is considerable difficulty in distinguishing ganders from geese, especially when young. Some experienced breeders determine the sexes by the difference in the voice, but that is a knowledge gained only by considerable acquaintance with geese. The form, size, length of neck, and size of the head, is some indication as they approach maturity, the gander being heavier, with a longer neck and larger head than the goose. A critical examination of each bird is a pretty sure method, but even this fails at times when made by a novice. On this subject Bailey says, 'Much difficulty is often experienced in selecting the sexes, and although practiced men are seldom mistaken, yet even they can lay down no rule that is easy to follow. Close examination may always be depended upon, but that is not easy to the uninitiated. There is a curious plan adopted in Cambridgeshire. All the geese are shut in a stable or a pig-stye; a small dog is then put in. It is said, and we believe with truth, the geese will all lift up their heads and go to the back of the place, while the ganders will lower and stretch out their necks, hissing all the time.'

"Before the days of steam or furnace heated houses and coiled spring mattresses, live geese feathers were a more important item and commanded a higher price than at present, and the fact that the breeding birds could be plucked from one to three or more times a season was an inducement to the keeping of geese which has very little force now. Comparatively few men pick any geese alive as in former days. The feathers obtained from the goslings fattened and killed for market are quite a source of income to the large dealer, as a good gosling will yield about enough feathers at present prices to pay the cost of picking.

"Geese are less liable to disease than any other domestic fowl, which, possibly, may account in some measure for their generally long life. Goslings well hatched are seldom lost, except through accident or exposure to hard storms while still very young.

"From the characteristics enumerated it is easily seen that the business of goose raising is of necessity somewhat restricted. It cannot be conducted in such a wholesale concentrated manner as is duck raising at the present time. The relatively large number of males required, the exclusiveness of the gander and his mates, the comparative large amount of range necessary for the breeding stock, and their aversion to close confinement, are some of the reasons why very large numbers cannot profitably be kept together. For the above reasons the business of goose breeding is never likely to be monopolized by a few breeders on a grand scale, but is likely always to remain in the hands of the many farmers who have low lying lands along brooks, rivers, and ponds, which, while comparatively worthless for other purposes, furnish ideal conditions for successful goose breeding. . . ."

The goose is still here and will be a thousand years from now, just as she has been from the beginning of human history; with the increase in the idea of a more general farming system, the flock of geese is fitting naturally and easily into the diversity of farming interest as one unit that will help to swell the cash balance at the year's end with little extra cost in feed, equipment or labor; there are a tremendous lot of folks interested in geese now and the number is steadily increasing.

As editor of *Cackle and Crow,* the writer during the past twenty-five years has had thousands of letters re-

American Buff Geese on James T. Eiswald's Little Falls Farm, Little Falls, N. J.

questing information on geese; how to feed, hatch, breed, market, house. Where to secure foundation stock and questions on breeds and varieties and their particular qualities and special uses; and inquiries multiply year after year.

There is little or no literature to which such inquirers may be referred. The United States Department of Agriculture issues a slim little bulletin on geese—authentic as far as it goes, but it certainly doesn't go very far. Then, Lamon and Slocum have written a book, now out of print, *Ducks and Geese*. It too, is excellent as far as it goes and it does go quite a long way on ducks but a very short distance on geese. Robinson's *Raising Ducks and Geese for Profit and Pleasure*, now long out of print, is also greatly over balanced by ducks and then there are fragmentary references to geese as sort of tail-end space-fillers in a number of books on poultry and that is about the extent of Goose literature in this country as far as I have been able to learn.

So, with the aid of some of my friends who know much more on certain angles of the subject than I, this book on geese is written; to more fully answer some of the thousands of questions which have been asked me in the past and perhaps forestall many more in the future that may be inspired by the steadily growing interest in the Goose.

I have sat by the hour and admired the lovely lines of the Chinese goose in water; handsomer than the Swan either on land or water and as highly intelligent as the Swan is deadly dumb.

Thousands of miles of travel and many days have been given up to an intriguing hunt for the well nigh extinct Pilgrim goose with its exceedingly interesting sex-limited plumage-color and its wealth of tradition as the breed accompanying our Pilgrim fathers on their voyage to the New World more than 300 years ago; with the satisfaction of seeing this valuable variety being bred by an increasing number of people and headed back to popularity from its definite start down the road to extinction.

To my friend, Oscar Grow of Atlanta, Missouri, I owe my first knowledge of this variety, gained through an article from his pen contributed to *Cackle and Crow* in 1934 and to him I also owe thanks for much interesting correspondence on our mutual interest, the goose.

I have stood on many different occasions near the best exhibits of African Geese in the world, staged annually at the Royal Winter Fair and Canadian National Exposition at Toronto and judged by my friend Andrew Taylor of the Government Experimental Farm at Ottawa, one of the few real authorities on Geese, and listened to the deep, melodious, organlike undertone of the Africans; unlike any other sound I have ever heard; and no other variety has this peculiar, deep, resonant melody in its voice.

I've had keen pleasure from long and interesting letters on geese from my friend Franklane L. Sewell, leading poultry artist of his day and from Harold Tesch who loves geese and breeds perhaps the best Toulouse in the world.

With Jack Deeter of Millbury, Mass., a fine fancier and finer friend, Clarence L. Sibley who has owned more varieties of wild geese and knows more about them than any man in America, Captain J. I. Lawrence of New York, Judge Ira Lloyd Letts who has the finest flock of Pilgrims in the country and many more, priceless friendships have been made and cemented through a mutual interest in this "forgotten bird" of the domestic poultry world.

I have exhibited all varieties of geese at many different shows and have had the thrill of winning the blue and even on occasion the supreme honor of "best in show." I have judged Geese at the largest shows in the country and have had a chance to learn and appreciate the sportsmanship of exhibitors of Geese.

I've raised Toulouse, Emdens, Pilgrims, Chinese, Crested Romans, American Buffs, Africans and Sebastopols and showing under good judges at leading shows have had the fun of competition with the same good sportsmen.

I've sat in the sun on May days and seen the little, downy day-olds come from the shelter of the mother's breast and take the first meal of tender, green grass and realized that they needed nothing further for food to grow and develop—if there could always be enough of this. No other domestic fowl can live from shell to mature stature solely on the commonest verdure on God's green earth.

Greatly, geese have added to my joy of living, in many different ways and I want to pass some of it along in this book, by making more knowledge of the goose available to the world.

"A team of twenty geese, a snow white train!
Fed near the limpid brook with golden grain
Amuse my pensive hours."
                                        —*Pope.*

CHAPTER II

Breeds of Geese

Lightweight Varieties

The Chinese Goose

Whether the Chinese goose is a direct descendant of the Grey Lag but bred to a different Standard from all our other varieties of geese which are said to have descended from that wild breed or whether the Chinese can claim as its ancestor the wild Chinese (Anser Cygnoides) which it very much more resembles, is a question which will probably never be answered.

That it may be of the same species as our other domestic geese is indicated by the fact that the Chinese breeds freely with them all and the offspring reproduces as freely. On the other hand the distinct and striking difference in body shape between the Chinese and the other kinds of geese, with the exception of the African, and that these two different types have existed from the time geese were first written about and that the Chinese is described back in the 18th century exactly as it is today together with its close resemblance to the Wild Chinese goose is strong evidence if not proof that it, rather than the Grey Lag, is its progenitor: furthermore, from all we can learn of poultry breeding in the ancient days, there was little, if any, selection for type and especially in a race of fowls bred almost solely for meat and feathers as the goose was, it seems very unlikely that rigid selection for countless generations such as it would have taken to fix the distinct body and head type we have in the Chinese goose would ever have been carried on.

31

"Lovely on a Lake" White Chinese. Property of Edw. D. Price, Wantagh, L. I, N. Y.

It seems much more likely to the writer that the domestic Chinese is simply the domestic descendant of the Wild Chinese of Asia which it so closely resembles in its present state.

I am inclined to agree with the Reverend Edmund Saul Dixon with reference to the common idea that all domestic fowl of the same species must be descended from the same common ancestor. He, in his book *Ornamental and Domestic Poultry* 1848 quotes from Jenyns' *Manual of Verterbrate Animals*, page 222, as follows: "The origin of the domestic goose is indeed unknown if we look to man or his influence to have originated so valuable and peculiar a species: but not unknown if we believe it to have been created by the same Almighty Power who animated the Mammoth, the Plesiosaurus, the Dinornis and the Dodo. For let us grant that the Grey-Legged goose is the most probable *existing* parent of the domestic sort: now, even that is becoming a rare bird and the more scarce a creature is in the wild state, the scarcer it is still likely to become. Suppose the Grey-Legged goose extinct; by no means an impossibility. Then those who *must* have a wild origin from which to derive all our domestic animals would be compelled to fall back on some other species still less probable. It is surely a simpler theory to suppose that creatures that were contemporary with the Mammoth have, like it, disappeared from the earth in their wild state, but have survived as dependents of man, than to engage in attempts at reconciling incongruities and discrepancies, which after all cannot satisfy the mind but leave it in as doubtful a state as ever."

As in the case of the camel which is said to have no wild relative in the world at the present time, why is it not reasonable to suppose the goose which history shows us to have been under the protection and in the service of man since history began, is the descendent of some wild species long since perished from the face of the earth?

Again Dixon, referring to the Brown Chinese in the same work says ". . . the old writers call it the Guinea Goose, for the excellent reason, as Willoughby hints, that in his time it was the fashion to apply this epithet to everything of foreign or uncertain origin. Thus, what we at this day erroneously call the Muscovy duck was then called the Guinea duck . . . Bewick calls it the Swan goose. The tubercle at the base of the bill, the unusual length of neck, and its graceful carriage in the water, give it some claim to relationship with the aristocracy of lake and river.

"Cuvier (Griffiths edition) goes further and calls it at once Cygnus Sinensis, Chinese Swan, and says that this and the Canada Goose cannot be separated from the true swans. A goose, however, it decidedly is, as is clear from its terrestrial habits, its powerful bill, its thorny tongue and its diet of grass. And therefore we have determined to call it the China Goose, concluding that Cuvier is right about its home, and other authors about its goosehood. . . . If fed liberally with oats, boiled rice, etc., the China Goose will, in the spring, lay from twenty to thirty eggs before she begins to sit, and again in the autumn after her moult from ten to fifteen more. I have never observed any disposition to sit after the autumnal laying. They are usually very late with their broods, but will rear them well enough if they are allowed to take their own time and do it in their own manner. My China Goose has now (June 1848) laid thirty eggs without intimating any intention of sitting; but she has annually brought up a family for the last five years, and I doubt not she will again, this season. The eggs of the China Goose are somewhat less than those of the domestic kind, of a short, oval, with a smooth, thick shell, white, but slightly yellow at the smaller end. The goslings when first hatched are usually very strong. They are of a dirty green, like the color produced by mixing Indian ink and yellow ochre, with darker patches here and there. The legs and feet are lead color, but afterward turn to a dull red. If there is anything like good pasturage for them they require no further attention than what their parents will afford them. After a time, a little grain will strengthen and forward them. If well fed, they come to maturity very rapidly; in between three and four months from the time of their leaving the shell they will be full grown and ready for the spit. They do not bear being shut up so well as common geese, and therefore those destined for the table are better for profuse hand feeding. Their flesh is well flavored, short and tender. Their eggs are excellent for cooking purposes. I have heard complaints of their being a short lived species from good authority, and that the ganders at least do not last more than ten or a dozen years. I cannot verify the fact, as my own experience with these birds extends only to about six years but it is quite in opposition to the longevity ascribed to other geese.

"Hybrids between them and the common goose are prolific: The second and third cross is much prized by some farmers, particularly for ganders: in many flocks the blood of the China Goose may be traced by the more erect gait of the birds accompanied by a faint stripe down the back of the neck. With the White Fronted Goose they also breed freely."

That the White Chinese were also in England at that time (1848) appears from a later chapter in Dixon's book where he states: ". . . my attention was first directed to these singular birds by Mr. Alfred Whitaker of Beckington, Somerset. 'I wish you could have seen the variety or species, it is so far superior in every re-

spect to the Brown. . . . it is spotless, pure white, (a
very few gray feathers have since appeared) more
swan-like than the Brown variety, with a bright orange
colored bill, and a large orange colored knob at its base.
It is particularly beautiful either in, or out, of the
water, its neck being long, slender and gracefully
arched when swimming. It breeds three or four times
in a season but I was not successful with them. . . . I
believe my birds are still in the neighborhood as I lent
them to a farmer to try his luck with them . . . the bird
deserves to rank with the first class of ornamental
poultry and would be very prolific under favorable circumstances. You will see both varieties of Brown and
White China Geese on the water in St. James' Park.
My Geese were from imported parents and were
hatched on board ship from China.'

"On visiting town in May, 1848, my efforts to get
sight of any White China Geese were unavailing.
There were none left! There were none now left in St.
James' Park; there were not any in Surrey Gardens,
choice as that collection is; nor were any visible at any
of the places where poultry is offered for sale. The
Zoological Gardens had parted with their specimens in
consequence of being overstocked with other things . . .

"From this difficulty I was most kindly relieved by
receiving a pair of White China Geese through Mr.
Whitaker's means. They are larger than the Brown
China Geese, apparantly more terrestrial in their habits; the knob on the head is not only of greater proportions, but of a different shape . . ."

From these excellent descriptions it will be seen that
the Chinese goose has changed very little, if at all, in
the hundred years which have elapsed since Mr. Dixon's experience with them as what he said of them,
well describes them today.

Trio of White Chinese. Original drawing by Franklane L. Sewell.

The White Chinese Goose

The White Chinese is considered by many the most beautiful of all geese. Its origin is unknown as stated previously; some authorities placing it with the Grey Lag and some holding to the opinion that the Wild Chinese goose is its direct ancestor.

"Streamlined" describes its general shape. Finely modelled, compact, well rounded body, erect carriage and long, gracefully curved, slender neck with a round knob at the juncture of head, and bill which is in nice proportion to head and neck; the whole effect is symmetry, grace and a harmonious blending of curved lines. The eye is blue and knob, bill, legs and feet are orange.

The White Chinese is as beautiful on water as the Swan but, unlike the Swan which is ugly and awkward on land, the Chinese goose is as smart in appearance and as graceful on land as on the water.

Its carriage is erect and its actions brisk and soldier-like. A flock in motion moves with military precision and their "goose steps" are far more graceful than the military goose-step of European armies.

Chinese have been called "the Leghorns of the goose family." They are the best layers of any variety. Early hatched goslings, if well fed and reared, will lay in the Fall and they start laying earlier in the Winter, as a rule, than any other geese and continue well into the Summer if the eggs are removed and the geese "broken up" when showing signs of broodiness. They lay from twenty to fifty or sixty eggs after the first year and I have had them lay as high as a hundred in one year.

I know of no efforts to increase egg production from geese by selective breeding but am convinced that the Chinese goose could, in a few years, be developed to the point that egg production alone, for table purposes, would make them profitable.

Chinese are good sitters and excellent mothers. They are the most sensitive to any unusual activity around the farm or yard, of any breed. They make the very best of watchdogs as nothing can move around the place either night or day, without them sounding the warning gong of alarm. For this reason alone they are an asset to any farm or country homestead.

In an article on marketing Chinese geese, the Editor of *Canadian Poultry Review* states, "The experience has been that they are quickly snapped up on account of their very attractive appearance" and this is in line with my own experience. For the large family or for family gatherings the big geese are in demand, but with the small family, the small, plump body of the well fitted Chinese is very acceptable.

The weights are: Adult Gander 12 lbs.; Adult Goose 10 lbs.; Young Gander 10 lbs.; Young Goose 8 lbs.

The Brown Chinese Goose

The Brown Chinese is identical with the White in size, in shape and in its general appearance of trig, trim smartness.

In color it is a grey with a brown tinge on top of body; each feather is edged with lighter grey, contrasting very nicely with the main surface color and in good specimens in full feather giving the effect of light rings running around the body and down the sides. The darker top color shading off gradually to very light grey or nearly white on the underparts of the body, and the breast and front part of neck are very light fawn with a dark, rich brown stripe running down the back of neck from head to back.

A pair of Brown Chinese. Original drawing by Franklane L. Sewell.

Brown, in his *American Poultry Yard* published in 1859 says, "The prevailing color of the China Goose is a brown which has been aptly described as the color of wheat. The different shades are very harmoniously blended and are well relieved by the black tuberculated bill and the pure white of the abdomen. Their movements in the water are graceful and swanlike. It is delightful to see them on a fine day in Spring, lashing the water with their wings, diving and rolling over through mere fun and playing all sorts of antics."

As in the White, trim, nicely rounded bodies are highly desirable with long, slim, gracefully curved necks, symmetrical, round knobs and refined heads and beaks. Any approach to coarseness in body, head or neck is objectionable and should be severely penalized in the show room, as under any well qualified judge, it is.

Breeders should bear in mind that this is an egg producing goose and that over size is not desirable and if big geese are wanted, the Toulouse, Emden or African is each ready to fill that need.

The bill and knob of the Brown Chinese is ebony black and the legs and feet orange, although formerly there was a variety with black legs and feet; probably a heritage from the wild Chinese goose which it seems to me was probably the ancestor of the domesticated Chinese.

The first of these geese mentioned in this country was by Bement in his *American Poulterers Companion* published in 1846 where he says, "Some beautiful specimens of the (Brown China) were brought out from China by Fletcher Webster, Esq., and taken to the home farm at Marshfield, Mass., some ten or twelve years ago."

These were perhaps the geese the Honorable Daniel

Webster of Marshfield exhibited at the first Boston
Poultry Show in 1849 sometimes referred to as "Canada geese."

The weights are: Adult Gander 12 lbs.; Adult Goose
10 lbs.; Young Gander 10 lbs.; Young Goose 8 lbs.

Thinking to get at once all the gold the goose could give, he
killed it and opened it only to find,—nothing.
—*The Goose with the Golden Eggs, Æsop.*

CHAPTER III

Heavyweight Varieties

The African Goose

The African in appearance is the gentleman or aristocrat of the whole goose family: His every line and action shows breeding, dignity and poise. His voice is deep and musical and his movements stately and assured.

His plumage is smooth and nicely finished; he has size without coarseness and his body is well made, symmetrical and nicely balanced on sturdy, strong legs.

I have never seen folks who were not impressed with their first sight of a pair or flock of Africans.

The origin of the African is doubtful; some authorities ascribing his beginning to the Grey Lag to which the beginning of all breeds of domestic geese is commonly credited. I would rather think and it seems more logical, that the African and his cousins of a more refined and slender type, the Chinese, are the domesticated descendants of the wild Chinese goose which is more like them in appearance than the Grey Lag or Grey Legged as it was originally called.

Bement in his *American Poulterers Companion* (1846) writes of the African as follows: "This is the largest of the goose tribe which has fallen under our notice; it is of the size of the swan, and it often weighs twenty-five pounds. We have now in our possession, one pair which we purchased for a gentleman in North Carolina, which will weigh, in common ordinary condition, over twenty pounds each. We once owned a gander that weighed twenty-four pounds. They are a noble bird, quite ornamental about the premises and add

43

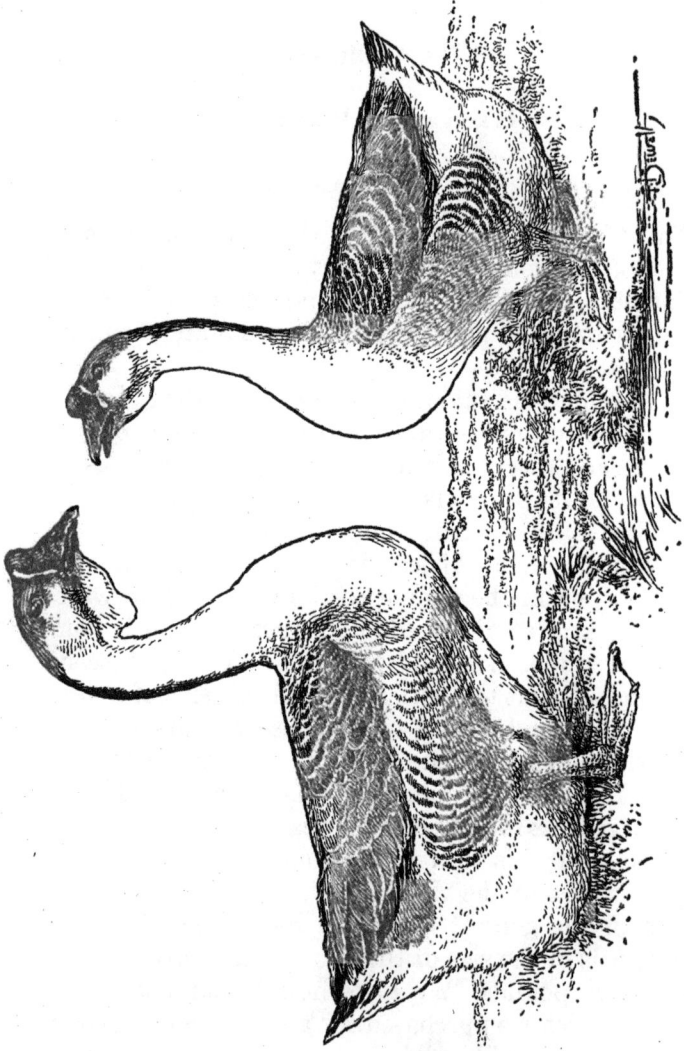

A fine pair of Africans. Original drawing by Franklane L. Sewell.

much to the scenery, particularly if a sheet of water be near. When floating on its surface they have a stately, majestic appearance, and in their movements they resemble the swan. They have a low, hollow, coarse voice unlike any other variety."

Dr. Bennett in his *Poultry Book* quotes John Giles who is said to have imported them under the name of "Guinea Goose" describing them as follows: "Brown-grey on the back, light-grey on the forefront, brown on the head and upper neck, prominent tubercle on the front of the bill, with a pouch, or dewlap under the throat. Weight will vary from twenty to twenty-five pounds each. Is a very rare and ornamental bird."

In *The Domestic Poultry Book* 1853 by T. B. Miner, editor of *Northern Farmer*, Rochester, N. Y., he publishes a letter from Mr. C. R. Belcher of East Randolph, Mass., in which he describes the African or the "Tchin Tchu" or China goose as he called it, very well indeed. A part of the letter follows: "As respects their properties, they grow to the weight of forty to fifty pounds per pair at mature size—say at two years of age. They are very productive; in fact, in this particular they excel all other varieties that I know. They commence laying very early in the spring season, and continue the production of eggs until late in the year. They are not at all erratic in their notions as to the place they deposit their eggs, but generally confine themselves to whatever locality may be prepared for them to lay in. They hatch three broods every year; but their eggs in the early part of the year are plentiful, and can be placed under other fowls to be hatched. They are hardy, thoroughly domesticated in their habits and have no requirements beyond what is common with other descriptions of geese. Various authors class them in the category of swans, and I think, they do so with degree of justification. Their majestic appearance when on the water, the peculiarity of their cry, and other features in their deportment and physical character, give preponderance to the idea that they are a species of swan. They are abundantly ornamental and, greatly prized in that capacity. They never have been in our markets for sale as food but those who have eaten them have been satisfied with their superiority for the table. Ultimately, however, they will become a very eligible article in the produce of the poultry raiser. At sixteen weeks old, goslings (or cygnets as the case

A great African Goose. "Champion Waterfowl," 98th Boston Poultry Show. Exhibited by Charles McClave, New London, Ohio.

may be) attain the weight of fourteen pounds dressed, and this by no means a rare thing."

Brown in his *Races of Domestic Poultry* is of the opinion that the African is the result of crossing the Chinese and the Toulouse stating as follows: ". . . It is essentially Chinese in character, varying principally in size and may be regarded as a descendant of the Brown Chinese goose but has probably been crossed with some other and larger variety, such as the Toulouse. Our opinion is that the African was originally the Brown Chinese, that it was taken to Africa by trading ships, and afterwards to America. Whether the crossing to secure increased size and thicker body took place in Africa or America we do not know; probably the latter."

To me it seems very likely that the African came in one of two ways. Either as an evolution from the Brown Chinese through generations of selection for heavier bodies and through rich and heavy feeding or as is pointed out by Mr. Brown, by the shorter and easier method of crossing with the Toulouse; and my guess would be the last. But we will always have that deep, sonorous, musical voice, like no other goose in the whole list, to wonder about, and so I think we will never know definitely whence came the African goose.

As to the "preponderance of the idea" that the African may be a species of swan, as stated by Mr. Belcher, I give little, if any, credence. The period of incubation is entirely different, that of the Swan being from one to two weeks longer and this in itself proves the African definitely belongs to the goose family rather than that of the Swan. But, whatever his ancestry and from wherever he came, the African is a beautiful bird and his beauty is not alone his claim to public favor.

In size, he is as large as the big White Emden. He is broad and deep and long but still finely proportioned

A pair of Standard Toulouse. Original drawing by Arthur O. Schilling.

and not cloddy and awkward. Africans are good layers, fine mothers, good grazers and easily fattened. They are rather quiet and are easily confined as they fly very little, if at all.

The African is colored much like the Brown Chinese; a greyish brown with each feather laced round the edge with a light grey, nearly white, and this body color shading off on the under body to nearly white. Down the back of the neck from the top of head to the back, extends a rich, brown stripe and the whole color scheme is rich and pleasing.

The African has a black bill and round knob at the front of the head at the base of the upper mandible; the legs and feet are reddish orange. Like the Toulouse, the African has a deep dewlap extending from the base of the lower mandible to neck. This is a marked characteristic and absence of dewlap disqualifies in the show-room. The dewlap is not known to serve any economic purpose but it is typical of the pure bred African and so is a "trademark" of an economically valuable member of the goose family.

The correct weights for the African are: Adult Gander 20 lbs.; Adult Goose 18 lbs.; Young Gander 16 lbs.; Young Goose 14 lbs.

The Toulouse Goose

The Toulouse is the most popular and widely bred of any variety of geese in this country and Canada. It is a development of the common grey goose of the farm-yard, reaching its highest state on the farms adjacent to the city of Toulouse in France from which it gets its name.

It is the largest of all domestic geese and its sturdy frame and big, rectangular, low-hung body, deep between short, strong legs, makes it the most admirable type for furnishing a great amount of flesh economically and quickly.

Of the origin of the Toulouse, Sir Edward Brown in his *Races of Domestic Poultry*, London, 1906 says: "It has already been shown the grey geese have been known since the species was domesticated and records are given of its existence in Britain for centuries. All the earlier writers from Gervasse Markham refer to it; but it is evident that, whilst the 'grey' has been used in the production of the Toulouse as seen today, other influences were brought to bear in securing an improved size, and there can be little doubt that this was obtained from the south of France. In the early part of the last century geese were largely bred in the Haute Garronne Department of that country, of which Toulouse is the capital, and even now the number kept there is greater than in any other French department save one. Dixon (in *Ornamental and Domestic Poultry, London* 1850) assumes that the Toulouse goose 'is only the common domestic, enlarged by early hatching, very liberal feeding during youth, fine climate and perhaps by age,' but this is not supported by the facts of the case, although the various influences named have each had their share in the evolution of our present stock. It would appear that in the fourth decade of the last century the then Earl of Derby, for his famous menagerie at Knowsley, imported some improved grey geese, as did others from Southern France. These were called at first, Mediterranean or Pyrenean, but finally the term Toulouse was adopted. From these, carefully selected, our present day stock has descended, and it is acknowledged that the English Toulouse is finer and larger than the French."

Of its history he says: "Importation of the improved Toulouse was immediately prior to the establishment of poultry shows, which later had a marked influence on the race, leading to the greater perfection of type and to increase of size. Whilst white geese are more general than grey among practical poultry-breeders, the size of the Toulouse has led to its use for crossing purposes. Upon the Continent, white geese are preferred but in America and the Colonies the Toulouse is largely bred."

Of the present status of the Toulouse in Britain, Reginald Appleyard, in his contemporary book *Geese* says: "This variety is of French origin, taking its name from the town of Toulouse but here again as in the Embden, English Toulouse have been bred to such perfection that they are now as a whole, far in advance of the average French birds. I make this without fear of contradiction as I have had many letters from French breeders who year after year import English birds . . . To anyone interested in breeding birds for special type and color, I would certainly say try this variety, it will give you plenty of scope to show your skill, plus the fact that you have a useful bird for utility purposes. Here let me say to the farmer or breeder who only requires birds for table purposes that the keel and gullet are of no use for table purposes, in fact a bird without them is, if anything, better. It is thus not necessary to pay special exhibition prices as birds which fail in these points will give just as good results as far as utility properties go."

Toulouse geese first came to this country shortly after 1850 and for a number of years did not catch the popular fancy of farmers in this country by whom, up to that time, geese were pretty generally kept, preferring a smaller and more active goose and only used the Toulouse now and then to improve the common stock. Toulouse were first exhibited at the Albany County, N. Y. Fair in the year 1856. The first American writers spoke disparagingly of them which tended to retard their popularity also. But as they gradually became better known and the interest in larger and more spectacular breeds of chickens developed, size in geese began to get the attention of farmers and fanciers alike and the Toulouse slowly started the long, slow climb to popularity in this country which ended in the position it holds today, as the most universally admired breed of geese in America.

Magnificent Toulouse. Government Experimental Farm. Ottawa, Canada.

John H. Robinson in 1924 wrote in *Ducks and Geese for Profit and Pleasure,* of the Toulouse, "For at least twenty years the Toulouse has been the most numerous of Standard breeds of geese in America, both in the heavyweight Standard type, and in the so-called Toulouse, which are in some cases of inferior size and in some cases grades of Toulouse, on common stock. Their relative popularity however, is not the same in all parts of the country. It is greatest in the central west where comparatively large flocks of Toulouse of good average standard quality may be seen on farms. Elsewhere fine geese of this breed are more likely to be found only in the hands of fanciers who breed them for exhibition, and to sell for exhibition, but by selling the smaller specimens to those who produce geese for market, contribute to the development to what may be called the common Toulouse Goose, which corresponds more or less closely with the farm type of the Toulouse in France. The result of such a division of the breed is equally unfavorable to the distribution of the two types. For no clear line of distinction can be drawn between them. Many of the stocks supposed to be pure Toulouse have in them a strain of other blood which accounts for the presence of characteristics disappointing to the buyer who supposes he is getting geese that are quiet, little disposed to roving or mischief, and of uniformly remarkable capacity for growth. On the other hand, the lack of a Standard for a variety of Toulouse goose of the popular size for a table goose leads to the distribution of much smaller geese under that name, and so gives seeming ground for the idea that the Toulouse goose is not greatly superior to the ordinary grey geese, and that specimens of Standard weight and over are freaks—abnormal specimens, like the occasional giants among human beings. It would greatly increase the popularity of the Toulouse goose in this country, if a farm market or utility type were standardized here, with the same relation to the exhibition type as in France."

With this last sentiment I cannot agree. Neither in the breeding of livestock or poultry of any kind can I discover the logic of breeding an economic variety to an exhibition standard which does not portray the ideal specimen from the economic viewpoint. Our poultry shows have been guilty of grave crimes against some of the most valuable breeds, not only of geese but of chickens, turkeys and ducks, by allowing the judges to place the prizes on specimens not of the greatest economic values simply because they excelled in some non-essential characteristic of shape or size or color solely, because that particular point was hard to attain. The Standard for Toulouse geese as set forth in the *American Standard of Perfection* is of an economic bird and if judges would make their decisions strictly by its terms, there would be no need of two Standards for the same breed as Mr. Robinson suggests.

The Toulouse gives one the impression that here is a bird to be depended on. Its massiveness and rugged contour, broad breast and long, wide back, deep keel touching the ground in good specimens, rather short, strong neck and short, rugged looking head and bill, all combine to impress one with the idea of tremendous strength, usefulness and the utmost in dependability.

The dewlap, or fold of skin that hangs under the head and throat from the bill to the upper part of the neck is a breed characteristic and denotes good breeding rather than having any particular economic value in itself. Like other breed characteristics in poultry and livestock which do not in themselves have any useful function, they are signs of breeding indicating the best specimens of the breed which has itself been bred and developed for some particular niche in its field and so these breed signs are the hall-marks of excellence in the various breeds which they help to identify.

A two foot fence will confine the well-bred Toulouse. Its big, heavy body makes flying almost impossible and its slow moving habits and contented disposition make it easily fattened for market and easy to handle and manage.

The upper part of the body and neck is a beautiful, dark grey with each feather on the back showing an edging of lighter grey, while the breast color shades off to the under body which is a very light grey or near white.

While size and weight are very important in Toulouse, these qualities should be combined with a smoothness of feather and a rounded and symmetrical outline rather than a shapeless, unattractive mass of fat and feathers. The back should be broad and slightly convex. The breast should be wide, full, deep and well rounded. The body should be carried almost perfectly horizontal in the goose and slightly higher in front in the case of the gander. The bill is reddish-orange and the legs the same color. The eye is dark hazel, bright, full and prominent.

Correct weights are: Adult Gander 26 lbs.; Adult Goose 20 lbs.; Young Gander 20 lbs.; Young Goose 16 lbs.

The Emden Goose

From the fact that the original importation of the Emden goose to this country was from the city of Bremen, the Emden was first called Bremen in the United States although in England it was known as Emden for the city in Northern Germany from which it originally came. Emden is now the approved name in this country and the variety is so listed by the American Poultry Association in the *Standard of Perfection*.

Bonington Moubray, Esq. in *Ornamental and Domestic Poultry* says of the Emden goose "At present, 1815, the Embden geese are in the highest esteem. They are all white, male and female, and of a superior, indeed very uncommon size. Whether or not, as might be expected, there be a counterfailing objecting in a corresponding whiteness, and thence defect of savoury flavor in the flesh I am unable to say, having as yet had no experience in the Embden variety of geese."

In passing it is interesting to note that he specifies the Emdens are "all white, both male and female" which all adds to the weight of evidence which is con-

A pair of Emden Geese. Original drawing by Arthur O. Schilling.

stantly growing that the common farm goose in England those days was more or less sex-linked in color and probably the progenitor of the present day Pilgrim. If there was not a variety of geese at the time Moubray wrote in which male and female differed in color why specify that the "Embdens" were all alike?

"To James Sisson of Warren, who exhibited three Bremen Geese of large size and beautiful white color, was awarded $3.00." This item in the list of premiums awarded at the Rhode Island Cattle Show at Pawtucket, R. I. and published in the *New England Farmer* October 20th, 1826 is the earliest mention I can find of any premiums paid on poultry in this country and thus far establishes the record for Mr. Sisson's geese as receiving the first prize-money ever awarded at a fair or poultry show on poultry of any kind.

In an edition of the same publication for September 2, 1825, there had appeared the following part of a letter reprinted from the *Portland Argus* written by Mr. Sisson to Mr. James Deering of Portland (perhaps the editor of the *Argus*) "... In the fall of 1820 I imported from Bremen (north of Germany) three full-blooded perfectly white geese. I have sold their progeny for three seasons. The first year at $15.00 the pair: the two succeeding at $12.00. Their properties are peculiar, they lay in February and set and hatch with more certainty than the common barnyard goose, will weigh nearly, and in some cases quite twice the weight, have double the quantity of feathers, never fly and are all of a beautiful snowy whiteness. I have been very successful this year in raising two flocks which I shall offer for sale in August: if you or any of your neighbors want a pair, I will sell them at $12.00 here and can send them at a trifling expense every week to land at Boston, so as to meet the Hallowell packet. Mr. Wm. B. Bradford, Jr., Grocer, No. 17 India St., head of the

Central Wharf will attend to it. I have sold them all over the interior of New York, two or three pairs in Virginia, as many in Baltimore, North Carolina, and Connecticut, and in several towns in the vicinity of Boston, and I wish to introduce them into the State of Maine.

"A line by mail directed to me at Warren, R. I. will meet with prompt attention."

"P.S. I have one flock, half-blooded that weigh on an average, when fat, 13 to 15 pounds. The full-blooded weigh 20 pounds."

This seems to establish Mr. Sisson, not only as the first exhibitor of poultry in America but as the first importer of Emden geese in this country.

Robinson in *Growing Ducks and Geese for Profit* says: "In the first American edition of Moubrays *Domestic Poultry* edited by Thomas G. Fessenden editor of the *New England Farmer* and published in Boston in 1832, the matter relating to the Emden was taken out and the following paragraph was substituted. 'A new breed of geese called Bremen geese has been introduced from Germany into the United States which we are told is decidedly superior to any before known in this country. They were first imported by Mr. James Sisson of Rhode Island, who received a premium from the Rhode Island Society for the Encouragement of Industry for the exhibition of this breed. They are said to possess the following advantages over any other animal of their kind: They grow to a greater size, may be raised with more facility, are fattened with less grain and make more delicious food.' In response to an inquiry from a subscriber as to where stock of Bremen geese could be obtained, and for more definite information in regard to the importation *The New England Farmer* of May 23, 1832 reprinted the

letter from Mr. Sisson quoted above: but, by a typo-
graphical error, or by confusion with the date they
were exhibited, the date of importation was given as
1826 instead of 1820. The consequence of this was
that when writers in the 'Hen Fever' period a score or
more years later, traced back the history of the Bremen
goose in America, they stopped at the record in 1832
and supposing it was correct, placed the date of the
Sisson importation in the fall of 1826. This was done
in *Bennett's Poultry Book*, 1850, where Col. Samuel
Jaques of Ten Hills Farm, Charlestown, Mass., was
also mentioned as an importer. The history of the
Jaques importation was then brought out, for the first
time, and appeared in the following letter, published in
the American edition of Dixon's *Ornamental and Domestic Poultry*, 1851:

Breeding Emdens. Government Experimental Farm,
Ottawa, Canada.

"Ten Hills Farm near Boston, Mass.,
December 12, 1850.

"J. J. Kerr. M.D.

"Dear Sir:

"My father, Col. Samuel Jaques, has had intimation from his friend Dr. Eben Wight of Boston that you are about to publish a book on the subject of Domestic Fowls, Birds, etc., and that you would be pleased to receive from my father some information relative to his Bremen geese—a name they have received in consequence of their having come from that place originally. I have my father's note to guide me in making the following statements, as well as his approbation that you should be furnished with them. In the winter of 1820, a gentleman, a stranger, made a brief call at my father's house: and, in conversation casually mentioned that, during his travels in the interior of Germany he had noticed a pure-white breed of geese of unusual size whose weight he supposed would not fall short much of 25 pounds each, provided they were well fed and managed. At that period a friend of my father's—the late Eben Rollins, Esq., of Boston—kept a correspondence with the house of Dallias and Company of Bremen and at his request Mr. Rollins ordered through that firm, and on my father's account, two ganders and four geese, of the breed mentioned by the strange gentleman. The geese arrived to order in Boston in the month of October in 1821; and I append a copy of "Directions relative to the geese from Bremen" given to the Captain of the ship in which they arrived. I hold the original in my possession, and transcribe it *verb et lit.*

'Emden 17 August, 1821

'The captain who is to take over these six geese will find the cages a little large; However, it is necessary that their lodgings be sufficient wide, if they shall arrive sound in America. Two geese which were sent to Bremen last year in a small box died on their arrival there; being water-birds, they want much more careful management then fowls; they ought to have constantly fresh water in abundance; a quantity of good sand and muscle shells serving for their digestion, must be in their feed box; there ought always to be sand and straw below in their cage for litter; also above the cage, as the birds perish otherwise by insects. The geese must be feeded; the geese used to pick straw from above down to the feet. The geese must be feeded with good clean oats, also with cabbage leaves.'

"Ever since my father imported the Bremen geese he has kept them pure and bred them so to a feather —no single instance having occurred in which the slightest deterioration of character could be observed. Invariably the produce has been of the purest white— the bill, legs and feet of the purest yellow. No solitary mark or spot has crept out on the plumage of any specimen to shame the true distinction they deserve of being a pure breed; like, with them, has always produced like."

Mr. Jaques' letter then goes on at length describing his father's care and management of the geese from the time he imported them in 1821 till the date of this letter written by his son in 1850 in a later paragraph of which he states— "It will be seen from what I have stated above, that my father was the original importer of this description and therefore is entitled to the credit of first introducing them to the United States. It is certain that he had the Bremen geese in his possession at least five years prior to the time when Mr. James Sisson of Rhode Island imported his . . ."

Colonel Samuel Jaques was one of the foremost breeders and authorities on poultry in his day. His Ten Hills Farm just outside of Boston in Charlestown, was noted for its great collection of both land and water-fowl and was the Mecca for those interested in poultry from all sections. The Colonel was "chairman of the Committee on Supervision" of the first Boston Poultry Show and was elected the first President of "The Society for the Improvement of Domestic Fowls" which was organized as a result of the interest aroused by this first show and to perpetuate the show. Incidentally, Colonel Jaques exhibited Emden geese at that first show.

This letter was written by his son 30 years after the first importations were made and this probably accounts for the mistake he makes when he claims his father was the first importer. It was probably made from records and the fact that the reprinting of the letter from Mr. Sisson in the *New England Farmer* eighteen years after it was first printed made the mistake of giving the date the Sisson geese were imported as 1826 instead of 1820 and that this had been quoted in several books published after the reprint was made, makes it very likely that Samuel Jaques, Jr., was wholly in good faith when he credits his father with the first importation. However, the records plainly show that to Mr. James Sisson of Rhode Island belongs the credit for introducing the Emden goose to the United States.

Emdens are beautiful birds. Large and symmetrical with snow white plumage, blue eyes and bright orange bills and legs. A flock makes a striking and attractive sight on farm or estate as well as a very useful and profitable livestock unit if properly managed.

The plumage is smoother and closer fitting than the Toulouse. The body is longer and not as low hung and deep. Keel not as pronounced and the habits and nature of the Emdens are more active and wide awake.

Unlike the Toulouse and Africans, typical specimens are minus the dewlap although sometimes specimens are seen having this non-characteristic feature. Specimens in shows on which dewlaps appear should be cut severely as it is evidence of Toulouse blood having been used by the breeder to increase size.

White feathers bring more on the market than colored and for this reason the Emdens are more desirable than the grey varieties, where raised in sufficient number to make the feathers a factor. They dress off beautifully as there are no dark pin feathers to mar the appearance of the carcass and this, as in all fowls, is an advantage when dressed for market.

As a rule, they do not lay quite as many eggs as the Toulouse or Africans but they are more active and there is apt to be less trouble from infertile eggs.

Weights are: Adult Gander 20 lbs.; Adult Goose 18 lbs.; Young Gander 18 lbs.; Young Goose 16 lbs.

A goose girl ermined is a goose girl still
And geese will gabble everywhere she goes.
　　　　　—Dorothy E. Reid—Not in Andersen.

CHAPTER IV

Medium Weight Varieties

The Crested Roman Goose

The Crested Roman is the rarest of all domesticated geese in America and probably the oldest.

Tradition traces it back through the ages to the alarm geese gave the Roman army on the approach of the Gauls 388 B.C., who, according to Homer: "—climbed the summit in such silence that they not only escaped the notice of the guards but did not even alarm the dogs, animals particularly watchful with regard to any noise by night.

"They were not, however, unperceived by some geese which, being sacred to Juno, the people had spared even in the present great scarcity of food, a circumstance to which they owed their preservation; for by the cackling of these creatures and the clapping of their wings, Marcus Manlius was aroused from sleep—a man of distinguished character in war who had been consul the third year before; and snatching up his arms and at the same time calling to others to do the same, he hastened to the spot where, while some ran about in confusion he, by a stroke of the boss of his shield, tumbled down a Gaul who had already found footing on the summit and thus started the defense which saved the Roman citadel."

Lucretious referring to the same event says, "The White goose, the preserver of the citadel of the descendants of Romulus perceives at a great distance, the odor of the human race."

Virgil, in alluding to the same occurrence ascribes the preservation of the capitol to a "Silver goose" and

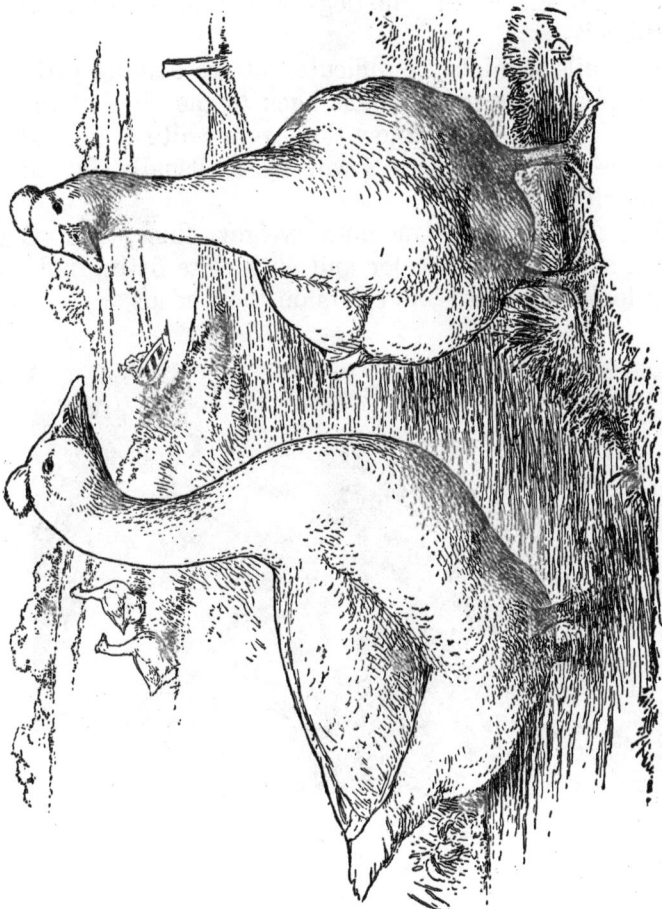

A pair of Crested Romans. Original drawing by Franklane L. Sewell.

Pliny, four hundred years afterward, says "the goose is carefully watchful; witness the defense of the capitol when the silence of the dogs would have betrayed nothing."

So, while it might be difficult to trace definite pedigrees of the Crested Romans back to the days of the glory of Rome, we do know they were white geese and that the earliest pictures of white geese usually showed them with crests.

The Romans are medium weight, finely shaped geese, snow white in color and with close fitting military looking helmet-like crests on their heads.

Crested Romans at ease. Golden Egg Goose Farms, New Haven, Conn.

They are very docile, fair layers and make excellent market carcasses. Their eyes are bright blue; bills, legs and feet reddish orange and their bodies sturdy and compact with rather rugged heads and necks and with keels not as fully developed as the Toulouse but noticeable especially in the females.

The weights are: Adult Gander 15 lbs.; Adult Goose 13 lbs.; Young Gander 13 lbs.; Young Goose 11 lbs.

The Roman Goose (Plain-head)

The Roman Goose, without crest, is probably the most popular and widely bred variety in England to-day. They are not bred in this country to any extent but would be a very useful addition to our domestic geese.

They are medium sized, smooth, plump-bodied, pure white geese, good layers and splendid foragers and easy keepers.

Sir Edward Brown in his *Races of Domestic Poultry* (London 1906) says regarding its origin— "Upon this point there is no direct evidence, but that is scarcely to be expected in a breed that has existed for twenty-three centuries.

"Among ancient writers many references are made to geese in Italy before Christ. Hehn says in *Wanderings of Plants and Animals* (London 1885) 'Among the Romans perfectly white geese were carefully selected and used for breeding, so that in course of time a perfectly white and tamer species was produced, which differed considerably from the grey wild goose and its descendants.'

"Homer, Livy, Lucretious and other ancient writers, mention both white and colored geese, and these birds were kept in the capitol at Rome as sacred to Juno, which, as in Egypt, meant that the priests retained them as long as possible for their own use.

"The conclusion come to is that the Roman goose was obtained by selection when its progenitor, the Grey Lag was captured and brought into the service of man.

"All the evidence goes to show that these birds are very precocious indeed, rapid in growth, and most pro-

A pair of American Buff Geese. Original drawing by Franklane L. Sewell.

lific layers, and we would suggest that the Roman would be a valuable breed both in England and America.

"A Belgian breeder who had tried the breed for a couple of years, writing October 28, 1899 (*Chasse et Peche,* November 5, 1899) says 'I have a Roman goose hatched May 10, which commenced to lay September 6 (then at the age of four months) and which, until today has laid twenty-two eggs—that is twenty-two eggs in forty-nine days.' This is borne out by the experience of others and it is stated that these birds will lay from sixty to a hundred and ten eggs from October to June.

"For the production of early goslings for which there is a good demand at profitable prices, such a breed should be valuable. The geese are very late sitters and other means of hatching should be adopted. The goslings grow rapidly and are quickly ready for the green gosling trade."

The body is rather long and broad; breast broad but not overly prominent. Back broad; neck long and rather slender and nicely arched. The head is fine and "breedy" looking; the eye is blue and prominent, surrounded with bare skin of pale yellow. The bill is rather short and thick, orange-red in color with a white tip. The wings are large but not carried as far back as some varieties. The legs are strong, reddish-grey in color with white toenails.

The general appearance is a sturdy, massive goose, smooth in outline with little if any keel and much smaller in size than the Toulouse or Emden.

Correct weights for the Roman are Adult Gander 15 pounds. Adult goose 13 pounds. Young gander 13 pounds. Young goose 11 pounds.

The American Buff Goose

From an origin as a general farm goose among the peasant farmers of Pomerania, the American Buff goose has, through forty years of selective breeding by poultrymen in this country, developed into one of the finest, most attractive and most uniform appearing of all our breeds of geese.

It is midway in size between the massive Toulouse and the smaller Chinese, Sebastopol and Pilgrim and it breeds exceedingly true to color and type.

Its shape is a happy medium between the tremendous depth and breadth of the Toulouse and the longer and more upstanding lines of the Emden and its color is distinct from any other variety.

A soft, greyish buff or fawn, lightly laced, with very light fawn around the edge of the feathers, shading off underneath the body to an extremely light fawn or nearly white, it makes one of the most appealing color patterns on any of our ducks or geese. There is nothing in the whole poultry world with exactly the same color shades and this alone makes it much admired by those who admire distinctive and unusual appearance in the creatures of their farms and homesteads.

The Buff Goose has a plump, deep body, smoothly feathered and trim, with a full breast and broad back. The female has a keel not as deep as the Toulouse but more pronounced than in the Chinese, Pilgrim or Emden and the male also usually shows some keel but less than the female. The Buffs are excellent layers, docile in temperament and make fine sitters and mothers. They dress off very nicely and are very satisfactory, all round business geese.

Sponsored by the American Waterfowl Association, the American Buff Goose was admitted to *The Stand-*

ard of Perfection by the American Poultry Association at its annual convention in St. Louis in 1946.

The eye is round, prominent, bright; and hazel in color. The bill and legs are orange and the whole color scheme harmonizes beautifully on this very useful and handsome member of the goose family.

The Buff is not as generally bred in this country as either the Toulouse, Emden, African or Chinese but it is increasing in popularity and for one who likes something just a little different from the common run, the American Buff will appeal.

Weights should be Adult Gander 18 lbs.; Adult Goose 16 lbs.; Young Gander 17 lbs.; Young Goose 14 lbs.

The Pilgrim Goose

There's a wealth of sentiment in the tradition surrounding the origin of the Pilgrim goose and its early history in this country.

Said to have been a part of the livestock with which the sturdy Pilgrim fathers provided themselves as seedstock for the homesteads and farms planned for their new life in the new and strange country, this interesting variety of geese came with their owners when the ship from the Netherlands landed on the bleak Massachusetts shore in 1620.

During the next two hundred years, they multiplied and spread with the early settlers over the length and breadth of New England as a part of the farm livestock so essential to the living of the early New England farmers.

Then began the trend toward specialization in Agriculture and geese became less common: the big breeds —Toulouse and Emdens came to us from Europe and attracted the folks still interested in our useful friend, the goose. The "Pilgrim" began to decrease steadily in

A trio of Pilgrim Geese. Original drawing by Franklane L. Sewell.

popularity and grow correspondingly less in numbers. By the end of the next century, the breed was practically extinct—and it was a pity.

For this small, sturdy, useful goose, besides the general hardiness and ability to rustle for itself, exceeded in this respect by no other variety, had the unique quality of sex-limited color and this no other breed has.

An article by Mr. Oscar Grow of Missouri, for years secretary of the American Waterfowl Association, an authority on geese and one of the best breeders of Toulouse geese and Rouen ducks in the country, written for *Cackle and Crow* about 1934 describing the Pilgrim goose, first aroused my interest in this unusual and at that time exceedingly rare breed. I had been interested in geese from boyhood and yet had never heard of the Pilgrim till then. Immediately I wrote Mr. Grow for more information; but except that tradition said they were very common in New England in early days, that he had taken the ancestors of his flock from Vermont 40 years before, that they were said to have been a part of the Pilgrim father's farm stock when they first touched Cape Cod and that there were very few in the country today, he had little or no information. But I must have some Pilgrims! I could not rest until I did. I finally secured several from Mr. Grow's small flock and later after much inquiry and after following many clues to flocks which invariably "had been disposed of some time ago" or gradually been displaced by other and larger breeds, I found the last remnant of a flock which had been bred by Mr. Leon Heck on "Lost Pond" in the wildest part of the Connecticut town with the record of having the largest area and the smallest population of any in the state; the town of Union, just on the northeast border of Connecticut on the Massachusetts line. The flock had diminished gradually during the thirty-five years during which Mr. Heck had bred them until there were just eight left. He had raised no young ones that year and seemed to have little interest in them. His interest however, seemed to grow as he discovered mine and at first he refused to part with any, notwithstanding my most persuasive line of coaxing and it was, I think, not until my third or fourth trip ("Lost Pond" was just about a hundred miles from my home) that I finally prevailed on him to sell me a pair.

Later I bought several more and shortly after, dogs got in the flock and destroyed what was left of this old family line. But of the origin I have thus far failed utterly to find any definite proof. None of the old writers have described them by the Pilgrim name or by any name. But from all the evidence I can find in a fairly complete reference library of old poultry literature which I have had fun collecting for more than forty years, I have come to the rather fixed opinion that the ancestor of the Pilgrim was nothing more or less than the common or farm goose of early Britain.

The following extracts from some of the more authoritative of the old writers, while none of them state definitely that the common goose of the old farms was sex-linked in color, do refer to the white ganders and grey geese; and there are many more which might be quoted. Rev. Edmund Saul Dixon in *Ornamental and Domestic Poultry,* London 1848 in writing of Pope's translation of Penelope's story of her dream:

> "A team of twenty geese
> A snow white train
> Fed near the limpid lake
> With golden grain
> Amuse my pensive hours."

says "Pope's version is both flat and inaccurate. 'The snow white train' I would bet Mr. Pope a dish of 'tey'

as he rhymes it—that Penelope's geese were not snow
white whatever the ganders might be."—And then
later on in the same book he quotes from Jenyn's *Ver-
tebrate Animals* published many years earlier as follows: "According to the Rev. L. Jenyn—'the domestic
goose is usually considered being derived from the
Grey Legged Goose, but such circumstance is rendered
highly improbable from the well known fact that the
common gander is *invariably white.*' And Peter Bos-
well of Greenlaw in *Poultry Yard,* 1841, says "In
order to have a good race of geese, the gander must be
chosen of a large size of fine *white,* and a lively eye;
the female either brown, ash or parti colored . . . The
color of the plumage ought to determine the color of
these birds. The parti colored are preferred to the grey
ones because the feathers sell much dearer; but the
latter are reckoned to be more fruitful and to give the
finest goslings."

And again later in the book he quotes M. Parmentier, a famous poultry writer of France as follows: "M.
Parmentier recommends the gander to be selected of
a large size of a fine white with a lively eye and an
active gait; while the breeding goose, he says, ought
to be brown, ash-grey or parti colored with broad
feet."

In Wright's *New Poultry Book* 1902, the Honorable
Sybil and Florence Amherst, famous English waterfowl breeders of those days, say: "Emden goslings
when hatched are yellow, some however with grey
tinge on their down. Observation has proved that the
grey ones are geese and the bright yellow invariably
ganders. It is said that Pfannenschmidt, who was a
well known merchant and connoisseur of geese at Emden, also certifies this; but this theory is not accepted
by all."

If this is so, it would indicate a natural tendency

toward this sex-color linkage of geese of white parentage and it will be very interesting for Emden breeders today to note whether this is still so.

I know with my own Emdens there is a difference in the down color when hatched but have never marked the day olds to see if this is a sex-linked characteristic.

C. N. Bement, author of the *American Poulterers Companion* 1856, says "According to popular opinion, the common goose is usually considered as having descended from the Grey Legged goose but such circumstance is rendered highly improbable from the well known fact that the common gander, after attaining a certain age, is invariably white."

And so, all through old poultry literature there are references like the above to white ganders and grey geese, making me believe that the Pilgrim goose, now one of the rare varieties, was at one time the most universally bred of all geese and was in fact, the old common goose of the British farmer.

The Pilgrim gander is invariably white with occasional specimens showing some partly grey feathers in wings during the first year and the goose a soft, French grey, shading off to white neck and head after the second moult. The eyes of the gander are blue and of the goose, dark hazel.

If you have ever tried to determine the sex of geese under a year old, you will realize the economic value of this sex-limited color characteristic. The sex of the goslings can be definitely told at a day old. The male is a creamy white and the female sort of olive or greenish grey.

The Pilgrim was admitted to the *American Standard of Perfection* in its 1940 revision and is sponsored by the American Society of the Pilgrim Goose, charged with the responsibility of bringing it back to the popularity it deserves.

Weights should be: Adult gander 14 lbs.; Adult goose 13 lbs.; Young gander 12 lbs.; Young goose 10 lbs.

The Sebastopol Goose

The Sebastopol goose which Lewis Wright referred to fifty years ago as being "not uncommon on the Danube" is another breed whose ancestry is in doubt.

Whence came its long, wavy, plumelike back feathering and beautiful curled feathers of the breast, so conspicuous on good specimens? Whether this unique and striking feather quality was in the beginning a mutation and has been fixed by selective breeding, or the Sebastopol is a distinct breed tracing back to some wild ancestor now long since extinct, probably no man can say.

At any rate, this is a striking and very appealing member of the goose family. Snow white in color with blue eyes and bright orange bill and legs and long, curved plumes hanging from back and sides, sometimes reaching the ground, and the short feathers of the front and under body curled closely, the Sebastopol is a most attractive bird, especially if one has water in which he may swim. On a lake or pond, Sebastopols are lovely.

They are quiet and domestic by nature; fair layers and good table fowl and a two foot fence will keep them within bounds as they are totally lacking in the ability to fly, from the peculiar construction of the flight feathers in their wings; the web of the feather being twisted and disconnected instead of smooth, connected and flat to offer resistance to the air, as in other birds that fly.

Sebastopols should not be kept with flocks of other geese if their chief beauty, their soft, plumelike feathering is valued. If crowded in flocks with many others,

A pair of Sebastopol Geese. Original drawing by Franklane L. Sewell.

these become broken and soiled and thus their principal beauty is destroyed.

If only a few other geese are kept and the Sebastopols are mated, they will usually hang by themselves and then there is no objection in running them with other varieties.

The Sebastopol is a small goose, easily raised and a good grazer. No country place with available water should fail to have a small flock of Sebastopols. They will add character, beauty and an individuality to the landscape and be a sight that most visitors will never have seen and will not soon forget. They will make far less trouble than Game birds or wild fowl of any kind and give as much or greater pleasure to the owner.

The body of the Sebastopol is short and plump; dressing off nicely. Their eyes are blue and the bill and legs orange.

Standard weights are Adult Gander 14 lbs.; Adult Goose 12 lbs.; Young Gander 12 lbs.; Young Goose 10 lbs.

Spring rides no horses down the hill,
But comes on foot, a goose girl still.
And all the loveliest things there be
Come simply, so it seems to me.
If ever I said, in grief or pride,
I tired of honest things, I lied.
—*The Goose Girl, Edna St. Vincent Millay.*

CHAPTER V

Rare and Ornamental Varieties

Russian Fighting Goose

Cock fighting has been carried on as a sport since the beginning of history and by men of all nations and stations but Gander fighting has been confined to one race of people so far as my research is able to discover.

For hundreds of years the Russians of both high and low degree have bred geese for the sole purpose of fighting. They have developed three distinct breeds or varieties and the sport has been popular especially among the well to do and the aristocracy.

Here is what Sir Edward Brown in his *Races of Domestic Poultry* (London 1906) says of these Russian Fighting Geese—"The first English reference we have been able to trace as to the fighting geese of Russia, is in the eighth edition of Moubray's on poultry (published in London in 1842) in which he states that at St. Petersburg they have no cock-pits, but they have a goose-pit, where in the Spring they fight ganders, trained to the sport and so peck at each others shoulders till they draw blood. These ganders have been sold as high as 500 rubles each. 'A short reference was made in work (Les Oiseaux de Basse-Cour published in Paris in 1895) to the Tula Goose under the term Race de Combat—' Our acquaintance with these birds which are the most striking in individual character of the species, began in 1899 at the exhibition held that year at St. Petersburg.

"We have been unable to obtain any information as to their origin, and until the study of poultry in Russia has advanced considerably must remain content with leaving the question alone.

"M. Houdekow says (*Traveaux du Congress Internationale d'Agriculture 1899 St. Petersburg 1901*) 'The only established fact is that this race has been bred in Russia from a very remote period, and that they have preserved their typical and fixed character.' The conclusion is that the type is entirely artificial and is not the result of natural selection but of man's choice for his own sport.

"These fighting geese, by which is meant the Tula and Arsanus, are bred in Mid-Russia to the south of Moscow and in the neighborhood of the upper Volga river. Formerly they were kept to a greater extent than is now the case, as goose fighting was a recognized sport and the breeding of first class specimens was a profitable pursuit, as was Game Fowl raising in Britain until sixty years ago.

"With the prohibition of the sport, although there is a great amount of goose fighting carried on surreptitiously in Russia, the reason for breeding such birds was removed, and less general attention was paid to them. Hence their place has to some extent been taken by other races, but they are still carefully bred in a more limited number of hands whilst their influence is seen in many directions.

"So far as is known, this class of goose is not to be met with in any other part of the world, is peculiarly Russian in character, and as already stated, is artificial in its special characters, for altho ganders are often good fighters, they are not naturally disposed to combat except in defense of their mates or their young.

"It is said that these birds are as keen for battle as are Game cocks and that their mates will encourage them to redoubled efforts.

"Keeping in view that these races have been espe-

cially bred for fighting, it will be understood that
weight of body and great strength of muscle are essential to success. Hence in developing the wings to enable birds to strike their opponents heavily, the breast
muscles must be increased accordingly, and thus, as in
the case of the Game fowl, the table properties were
enhanced although there was no direct intention of
securing the result.

"M. Houdekow again refers to the fighting breeds as
follows: 'Our geese are indebted for their existence to
the sport of goose-fighting; Nearly all our races of
geese are fighters. The principal quality which has
secured their appreciation is the faculty of fighting
each other furiously; until one of the combatants is
dead the other will not leave the field of battle.

" 'These geese are not very productive. Thirteen
eggs per annum is the usual record, but sometimes
birds produce twenty eggs. These geese have remarkable endurance and do not require any special feed or
care. The flesh is of an excellent quality, is not fibrous,
and is very fine in flavor. It is impossible to fatten
fighting geese to the same degree as the Toulouse or
Embden but their flesh has a good taste, more like that
of wild game. They are good setters, and very careful
of their young, defending them against all species of
birds and animals.'

"Our observations in Russia support these remarks,
except that for general purposes the flesh is hard and
they require to be well hung after killing before they
are cooked.

"Thus they are chiefly useful for crossing as in the
case of the Game and Indian Game fowls, with soft-
fleshed races, under which conditions the results are
seen in an abundance of breast meat of a very fine
quality. Difficulties must necessarily arise in the case
of birds in which the pugilistic nature is so highly de-

veloped due to their quarrelsome nature, but where single males are kept for crossing that would be reduced to a minimum. The use of the Russian geese with some of our Western breeds is well worth a trial for the development of meat qualities.

"All the races of Russian geese named are very similar in general characters.

"The body is long, broad and very deep, with a full, round and massive breast, and a wide, straight, flat back; the neck is of medium length, very strong and slightly curved near the head; the head is the peculiar feature of the breed, and entirely distinct from that of any other bird with which we are acquainted; it is short and nearly round, with a wide forehead and well developed cheek muscles, and in older birds there are often two knobs or protuberances, which increase in size with advancing age, on the upper part of the skull; the bill is very short indeed, and stout at the base, the line of the upper mandible being in line with the front of the head which together with the bill, is practically the same length as depth, in conformation resembling the head of a bull-dog.

"Beginning from the nostril, the surface of the bill is ribbed and the mouth, between the upper and lower mandibles has a strange rounded appearance, seeming even when closed as if it were partly open; the color of the bill is pale yellow with an ivory tip; the eye is large, full, and nearly black as a rule, but some specimens have grey or light blue pupils; the wings are very large and powerful, with strong shoulder muscles; the tail is short; the legs of medium length, very strong and well set apart, with large round feet and are in color dark orange-yellow.

"From the above it will be seen that these birds are built upon powerful lines, and they are close and compact in plumage and firm in flesh.

"It is evident that these three classes are all of the same race, and we have, therefore adopted the generic title of Russian for all.

"TULA—These birds are the best known, and are chiefly bred in the central districts of Tula, Kalonaga, Tamboff, Kizan, Koursk, Vladimir, Nijni, Nororod and upon the borders of Oka and the upper parts of the river Volga. They are generally grey in plumage but some are clay colored. Weight 13½ to 17 pounds.

"ARSAMAS—Pure, spotless white in plumage and larger in size of body, as they range from 17 to 22 pounds.

"KHOLMOGORY—This variety differs from the others in that they have a longer neck, and the bill is more natural, being of ordinary length; the upper line of bill is curved at the end, giving it a hooked appearance. They are also used for fighting, but not in the same extent as the Tula or Arsamas. This is the largest Russian breed of geese weighing from 20 to 23 pounds."

I find no record of any of these breeds ever having been imported to England or the United States and from the above description it would seem that here is a useful bit of promotion work for some of our educational institutions or for the American Waterfowl Association or even some private breeder with improvement of geese for meat qualities as an objective.

Just at this time the crossing of the Cornish fowl, referred to above by Mr. Brown as Indian Game, with our American varieties of chickens to improve placement and quality of meat is becoming very popular.

These Russian geese with the same heavy wing and breast muscles should have the same effect if crossed with our domestic breeds of geese.

As in pedigree breeding for eggs, here is a chance for

some really practical and useful breed improvement work.

The Canada Goose

For some unknown reason the American Poultry Association includes the Canada goose in the list of domestic geese for which it provides space and descriptions and formulates standards in its *American Standard of Perfection.*

It is not a domestic goose but in the interest of conformity and as exhibitions provide competitive classes for it, I shall give a description and provide space with the list of domesticated varieties of geese in this book.

The Canada is of a species distinct from the domestic varieties and while it will cross with them, the result is a hybrid which, like Mark Twain's mule, has no cause for "pride of birth or hope of posterity."

As previously noted, the Rhode Island goose raisers practiced crossing the Canada gander with domestic

100% hatched and raised. This pair of Canadians produced six eggs and raised six goslings. Government Experimental Farm, Ottawa, Canada.

geese for the production of the famous "mongrel goose" for market. At one time this bird was considered a great delicacy and sold on the Boston market at very high prices.

The Canada is a strikingly handsome variety and is hardy and very ornamental for parks or country estates. It is, however, a poor grazer and requires more feed than the ordinary domestic varieties and should be pinioned to ensure its "staying put" when the migratory instinct pulls for warmer climates in the fall of the year or cooler in the spring.

Pinioning consists of cutting off the last joint of one wing which makes the bird incapable of making long flights and is a very simple operation. By using a little care and a sharp knife, the joint may be found and the knife inserted between the two bones at the joint and hardly a trace of blood will be shed if the operation is expertly and deftly performed.

Canadas are shy layers, usually laying from four to eight eggs in a clutch which they will usually repeat if broken up when the first sitting fever starts.

As a rule the geese do not breed until the second year although sometimes the ganders will mate when a year old.

Gander and goose are alike in color. The head and neck is black with a sharply defined white marking on each side, running up behind the eyes and meeting under the throat.

The wings are grey with a slight pencilling of lighter grey across ends of feathers and the back is grey. The tail is black and the breast a soft grey shading off to a darker grey at legs and from there to tail an extremely light fawn or white. The bill, shanks and toes are black.

This goose is beautiful in contour and symmetry and its plumage is smooth, sleek and tight fitting. It is of a haughty and rather distant disposition and while not particularly pugnacious does not mix well with other geese.

Regular weights are: Adult Gander 12 lbs.; Adult Goose 10 lbs.; Young Gander 10 lbs.; Young Goose 8 lbs.

The Egyptian Goose

Like the Canada goose, the Egyptian is listed in the *American Standard of Perfection* along with many varieties of domestic land and water-fowl. It would be just as appropriate to so list the Pheasant, Quail or Partridge.

The Egyptian is essentially a wild bird in habits and disposition and needs to be pinioned (the first joint of the one wing removed) if the owner has any desire to keep it at home. They can fly like birds and will if given the opportunity.

It is unlike any other goose and seems to be more a wading bird than one designed for swimming. Its legs are long for its size and it is one of the smaller varieties.

This goose is abundant along the banks of the Nile and on the African continent generally. It also visits the southern shores of Europe and is often seen in Sicily. According to Temminck, it was held in veneration by the ancient Egyptians and it figures frequently on the monumental remains of that extraordinary nation.

Its plumage is lovely. A beautiful blending of soft, pastel shades emphasized by sections of brilliant coloring; there is I think no other goose that approaches it in rich, exotic hues.

The base of bill and space surrounding the eye is rich, chestnut brown; cheeks, crown and throat yellowish white. The neck is yellowish-brown in front

Left—Egyptian Gander. Right—Sebastopol Goose. (Photograph from the Bureau of Animal Industry, U. S. Department of Agriculture.)

and rich red-brown in rear; the upper part of back, the breast and flank light yellowish-brown minutely waved with a darker tint and the legs, shanks and feet reddish-yellow. The center of the breast and belly nearly white with a dark horse-shoe shaped patch of dark, rich brown where the two sections meet. Vent and under-tail coverts buff orange; the lower back, rump, upper-tail coverts and tail, black; wings as far as the greater coverts, pure white, the latter having a deep black bar near their tips. The main wing feathers chestnut red with greyish brown color on the inner webs; Secondaries black at the tips with the outer web a brilliant irridescent green.

The bill is reddish-purple and the eye a bright golden.

It is hard to describe the beauty and harmonious blending of these shades and colors and the effect is just as beautiful as the disposition of this bird is pugnacious and cruel.

Unquestionably he will dominate any other variety of duck or goose on the place with the possible exception of the Muscovy drake which is twice his size and fully as fearless and capable a warrior.

The Egyptians are rather poor layers, laying from four to seven eggs in each clutch and laying two clutches in a season if properly managed.

They do not breed until two years old and are a purely ornamental breed.

Weights should be Adult Gander 5½ lbs.; Adult Goose 5 lbs.; Young Gander 5 lbs.; Young Goose 4 lbs.

The Grey Lag Goose

According to popular belief the little, inconspicuous Grey Lag or as the old writers give it "Grey Legged" goose, is the Adam of all the domestic varieties of geese.

The origin of its name is in dispute among writers and authorities in ornithology and there is not too much unanimity of opinion as to its being the single ancestor of all the breeds of Geese.

Captain J. I. Lawrence in the *New York Sun* says: "By a curious coincidence, practically all the domestic poultry breeds, domestic ducks and domestic geese in the world stemmed originally from a single wild species of each group, though the world has many families and genera of wild land fowl, wild ducks and wild geese.

"And by another curious coincidence, the possible exceptions to the fact are narrowed down to one in each case. It is generally believed by naturalists that all domestic poultry breeds came from the red jungle fowl (Gallus gallus) of the pheasant family, though there are other jungle fowl with similar characteristics, and a host of other gallinaceous birds; and the single exception, by conjecture, is the Malay gamecock, sometimes three feet tall, whose origin is vaguely traced to some long-extinct giant jungle fowl, more fabulous than real.

"The domestic ducks in their legions of types almost as varied as those of poultry, are supposed to have come, one and all, from that universal species of wild duck, the Mallard; and the exception is the farmyard Muscovy duck, which is merely a tamed and domesticated Brazilian duck of similar type, color and characteristics, suspected of being more like a goose than a duck.

"The domestic geese, of many and varied types, acknowledge the wild grey lag goose of the Old World as their common patriarch; and the grey lag's habitat extends from the British Isles to China, but the exception is the tall, swan-necked China goose, with its cousin, the African, both of them probably descended from a wild Siberian species of a swanlike type.

"Among the novelties in last week's (1939) Boston Poultry Show were two wild grey lag geese from the collection of wild and domestic geese at Golden Egg Goose Farms owned by Paul P. Ives of New Haven. The species never has been domesticated in its natural form, and is a subject for the aviculturist rather than the farmer; but Mr. Ives is both naturalist and farmer, and he searched a long time for specimens for his large collection. He acquired the goose and gander seen at the show quite recently (from Mr. C. L. Sibley, noted authority on wild geese and ducks. Author) and will attempt to breed from them during the coming season.

"Many interpretations of the peculiar English name of the species have been offered, but it is generally believed that it was applied by peasants after they observed that the geese did not migrate with other wild geese, but 'lagged' behind, and often remained on the summer breeding grounds through the winter.

"The grey lag is grey on neck and back, but brownish on the wings, with bluish gray on the coverts. Breast and under parts are light grey shading out to white, the breast showing faint transverse barring, and the tail is brown with white margins. The bill is pale pink or yellow, with a white tip. On the neck the feathers are peculiarly curled or crimped, and the same effect is seen in the domestic Toulouse goose, and in some other varieties."

Bement, writing in the *American Poulterers Companion* in 1856 states: "According to popular opinion the 'common goose' is usually considered as having been derived from the 'Grey-Legged Goose,' but such a circumstance is rendered highly improbable from the well known fact that the common gander after attaining a certain age is *invariably white."*

Incidentally, this helps to substantiate the general belief that the Pilgrim was the common goose of early New England.

In this connection it is interesting to note that Selby in his *Illustration of British Ornithology* thus expresses himself regarding the "general stock of this country." *"The gander is usually white* or with a preponderance of that color while the geese have various shades of ash grey and a dull, leaden brown with it; a preference is often expressed for those that have no white whatsoever excepting only on the lower parts of the body." Selby also refers to the Grey Lag as the "Grey Legged Goose."

But, Grey Lag or Grey Legged, whether or no it is the ancestor of all domestic geese, it is an interesting variety and the generally accepted tradition that all our domestic geese evolved from it makes of it a "show piece" for any collection of water-fowl.

The Grey Lag is a European variety and in appearance is much like the common grey goose of the farms of England but smaller. It is colored like the Toulouse except for the grey color of the legs which in the Toulouse are orange.

Its weight is about four to five pounds and it breeds in confinement.

The American White Fronted Goose
(*Anser albifrons gambeli*)

(Grey Brant—Speckled-Belly—Prairie Brant)

The White Fronted goose is a slightly smaller, somewhat slimmer bird than the Snow Goose. The upper front part of the neck is a brownish grey with lighter edgings to the feathers. It has a white band at the upper base of beak and forehead backed by dull black.

The back as it approaches juncture of the tail is nearly white. Wings and tail dusky grey; sides greyish brown broken with darker color. Breast lighter than throat and marked like the white underbody with dark patches. The bill is pinkish or soft red. The feet are yellow and eyes are brown. Young birds lack the white front on head and are browner generally than the mature birds and also lack the dark patches on the underbody.

The favorite habitat for the winter vacation of the White Fronted is California and as far south as Cuba and they may be found during the winter months as far east as the midwest but are very rare on the Atlantic Coast. They summer in the far north where they nest and hatch their young, moving north about April and returning for the winter's sojourn between the fifteenth of September and the first of November.

The voice of the White Fronted goose is higher pitched than the melodious organ-like honk of the Canada with a sort of quickly repeated cackle-like ha, ha, ha, as they fly swiftly with forward thrusting heads, from their early morning feeding grounds on some stubblefield to their daytime rendezvous on some open lake or wide river.

This journey is made twice a day. Once at daylight to the inland feeding ground and once just before night to the same feeding place, on some low swampy marsh or wooded tract, stubblefield or wide spread prairie where there is grass; and seeds and roots near the surface, which they love.

Like other geese the White Fronted Goose is strictly vegetarian and when seen with tails up and heads deep in the water they are grubbing roots and eel grass from the bottom instead of shellfish or other animal life as is sometimes supposed.

The White Fronted, grand game bird as he is, like other varieties of wild geese is steadily growing less and becoming scarcer year by year. More is the pity.

The Snow Goose
(*Chen Hyperborea Rivalis*)

(Also Called White Brant; Wavey; Blue Winged Goose)

This attractive wild goose is entirely white except for the flight feathers of the wings which are dull black and the wing coverts which are a light grey. They are about two thirds to three quarters the size of the Canada goose and are very rare in the eastern part of the United States north of Virginia.

The bill is carmine and the legs a dull red. Young birds have feathers on upper part of body greyish with white edging.

Snow geese are natives of North America. During the winter they may be seen in the southern, central and Pacific coast regions, migrating to the far north for their breeding and nesting grounds in April and returning to the United States and as far south as Cuba about October.

Neltje Blanchon delightfully pictures them with words in the book *Birds That Hunt and Are Hunted* as follows—"The dullest imagination cannot but be quickened by the sight of a great flock of these magnificent birds streaming across the blue of an October sky like a trail of fleecy white clouds. Such a sight is rare indeed to people on the Atlantic coast north of the Chesapeake; but in the Mississippi valley during the migrations, on the great plains and in parts of California all winter, fields are whitened by them as by a sudden fall of snow. Lakes in Minnesota may still be

Pair of Snow Geese with goslings. Flock of Snow Geese. Pair of
Snow Geese with nesting. Government Experimental Farm,
Ottawa, Canada.

seen reflecting their glistening whiteness as if snow peaks were mirrored there; and in the Sacramento and San Joaquin valleys, in Oregon and beyond, they are still sufficiently abundant to be hunted on horseback by the indignant farmers, who see no beauty in their plumage to compensate for their devastated fields of wheat that the hungry flock nip off close to the ground.

"But like most other choice game birds, the snow goose is fast disappearing. Who that knows how fast this decrease is, ever expects to see such flocks as gladdened the hearts of Lewis and Clarke when they reached the mouth of the Oregon?

"Closely associated with the White Fronted and the Canada geese, Snow Geese may be named, even when too high up in the sky at the twilight of dawn or of evening for us to see its dark tipped wings and white plumage, by the higher pitched, noisier cackling that distinguishes its voice from the Laughing Goose and the mellow honk of the Canada. It migrates by night and day; observes punctual meal hours like the rest of its kin; keeps a sentinal always on guard while it feeds in the grain fields or roots among the tidewater flats and grassy patches bordering streams; circles, gyrates, tumbles and floats above the water on returning from its feeding grounds, in short it behaves quite as other geese do when intoxicated with food."

The Lessor Snow goose is practically a smaller edition of the Snow; is identical in plumage and resembles it closely in its habits.

It nests in Alaska and confines its roaming in Winter pretty generally to the southwestern section of the United States. By pinioning, these birds may be bred in confinement and become as tame as the domesticated varieties with a little attention. A flock of Snow geese is an ornament and a perpetual delight on the farm or estate of one who loves our feathered friends.

Snow Gander and Chinese Goose with hybrid progeny at hatching and in October. Government Experimental Farm, Ottawa, Canada.

Sturdy African breeders of excellent type. Government Experimental Farm, Ottawa, Canada.

GB-249

CHAPTER VI

Selecting Geese for Breeding

Strong, vigorous specimens should always be selected for breeding. Health, vigor and activity are to be looked for always in picking out the specimens which we wish to become the parents of our annual crop of goslings.

Size is essential. Breed-type and correct coloring is desirable but without health, strength and virility in the parent stock, your breeding experience will be just as disappointing with geese as with other poultry or livestock lacking these essentials.

Select then, strong, vigorous males with robust, masculine looking heads, alert, bold appearing eyes. Birds not too large and coarse but not undersized. A bold, courageous looking and acting male will as a rule prove the best breeder. Courage seems to be an attribute of virile specimens in all birds and animals.

In color (with the exception of Pilgrims) as in general form, the female in all varieties is much like the gander; except for a somewhat deeper body and keel and a more feminine appearing head and usually somewhat smaller in size. National Encyclopedia says, "The true geese are distinguishable from ducks by their usually greater size, longer necks and by the male and female being alike in color" which just shows that the author who did the piece on geese didn't know all of his subject. The Pilgrim Goose is dove-grey and the gander is invariably white, with an occasional specimen which shows a little grey in wings when young. The sex of the Pilgrim goslings may be easily told at hatching by the color; the male being a yellowish-orange and the female a sort of greenish grey.

Especially for the novice it is difficult to determine sex. In all varieties except the Pilgrims, the color pattern is identical as is the feather growth; and the body shape is very much alike in all breeds.

To the experienced eye, the sex can generally be told by observation, however, as there is usually some difference in the size of the goose and the gander; the goose as a rule is smaller and shows more "keel"; the gander being more upright in carriage, has a more masculine appearing head and eye and the voice is high and shrill while in the goose it is lower and deeper in tone. This refers to the tones used in ordinary "conversation" in the flock. When alarmed or excited the voice is much alike in both sexes.

In geese under a year old, however, even the experienced breeder is at times uncertain about the sex and there is but one absolutely sure way of telling the male from the female. This is called "sexing" and is done as follows: Have an assistant hold the bird upside down with back to him and breast toward you. He should hold a thigh and wing in each hand. Now place the finger tips of one hand around one side of the vent and of the other hand around the other side of the vent. Gently press them into the body and at the same time smooth open the vent with the thumbs. In the gander the exposed surface will be smooth and the penis or organ will be exposed by pressure. In the female, the exposed surface will be wrinkled and loose.

In selecting geese for breeding, health and vigor are of course equally essential in male and female. Good, smooth, clean looking plumage is generally a sign of good health and birds that are active and wide awake appearing usually are vigorous and in good breeding condition. After the signs that indicate vigor and vitality, look for breed-type and color. Bear in mind that breed-type in all livestock has been developed by breeders with a view to the greatest usefulness for the particular object or use for which the breed is designed.

Thus, the strong head, heavy neck and deep low-hung body with broad back and breast of the Toulouse indicates the type best suited to put on a large amount of good, solid flesh economically and quickly. The Emden, with its more active habits and somewhat smaller size, is built for a more active life and more ability and inclination to rustle for itself. On the other hand, the Chinese varieties are alert and active, with bodies higher from the ground and much smaller in size but plenty big enough to most economically produce more eggs than any other goose; thus each variety for its special purpose in life.

So, with a foundation of health, vigor and virility to start with, add the secondary consideration of breed-type and to this, the correct colors and shades of color and you have the design of the best breeding birds.

Always remember that "standard" or exhibition points are all based on usefulness plus beauty and both are essential if you are to produce the finest in the geese you are breeding; just as both are essential in all things connected with the life you live if you get the most out of it.

The best livestock men in all lines, whether it be Beef or Dairy Cattle, Sheep, Swine, Chickens, Ducks, or Geese, consider uniformity of breeding resulting in uniformity of size and type of considerable economic importance.

It is obvious that if all specimens in a herd or flock are of uniform size, shape and temperament, resulting from uniform breeding to a certain standard, they will respond to the same system of feeding and management in somewhere near the same degree; what is right for one will be approximately suitable for all, whereas if the herd or flock consists of carelessly bred individuals of many different sizes, temperaments and blood lines, the feeding method that is correct for one may be wrong for another and so some will not respond equally with others and the maximum profit from the group will not be attained.

"Alexander" and his mate with eight youngsters. "Alexander" was Champion Waterfowl at 96th Boston Poultry Show. Owned by Edw. D. Price, Wantagh, L. I., N. Y.

Furthermore, uniformity in carcasses, especially in poultry, is an economic asset when marketed and will bring a higher price than those of varying shape, size and color.

Feeding for Breeding

Feeding for breeding is one of the most difficult problems connected with goose raising. To have the goose in a condition to produce plenty of eggs requires an abundance of feed before the laying season comes on. Fat is the greatest obstacle to fertility and plenty of feed is apt to make too much fat. Especially true is this of the gander. It's over-fat ganders that are most apt to be sterile.

The problem is greatest with show geese. Size and weight count heavily in the show pen. Of the large varieties, most judges give these points altogether too much value in judging. The *Standard of Perfection* gives them no justification for over-emphasizing size and weight but the fact remains that the big, over-fat birds usually win. " 'Tis true 'tis pity and pity 'tis, 'tis true."

This means that the geese shown are generally way over-fat and as a rule are very uncertain for breeding. To make the weight suit most judges they must be so fattened that their breeding value, for that year at least, is practically ruined. Sometimes they are value-less as breeders for life.

From the time the breeding geese are started on grain in the Fall, they should have plenty of roughage. Clover, second or third crop alfalfa or other fine cured hay with good bright corn stalks, make excellent bulky food. If possible, they should have green food every day in addition to dry roughage. Cabbage, beets, waste apples; in fact any discarded vegetables or fruits make a valuable part of their diet; but root vegetables should be chopped or sliced.

Geese kept for breeding should have little corn. Give them a good mash of ground oats, wheat bran and corn meal, in about equal parts, once a day and all the roughage and green food they want and they should come into the breeding season in the proper condition to produce a good number of eggs and not be so fat as to have the eggs infertile.

If you live within easy shipping distance of New York, plan to have your geese lay early. This can be managed by starting to feed heavily early in the Fall. Plan to market the first eggs in New York City before

Easter when there is always a good market at a good price; $2.50 to $3.50 per dozen on the wholesale market. Then, after Easter, set the eggs; this will bring the young goslings along about the time the grass begins to get green and nutritious.

If you have no market for the eggs early in the season, plan to have your geese come into production later, as nothing is as good and as economical for growing the goslings as tender, green grass.

Breeding

The goose is often monogamous and in the wild state is always so; when once mated, it is generally for life. As a rule, the domestic gander is faithful to his mate or mates; for sometimes he departs from the natural order of his kind and takes unto himself two wives and in some cases three or four but whether he takes himself one, two or three mates, to these he is mated for life in the great majority of cases.

Dixon says: ". . . It might be urged that the Domestic Goose is polygamous, whereas all Wild Geese that we are acquainted with are monogamous. It is true that Wild Geese, in captivity, will couple with the females of other species; but that takes place by their utterly neglecting their own mate for the time, not by entertaining two or more mistresses at once. It will be replied, that habits of polygamy are the effects of domestication; but what proof have we of such an assertion? Domestication has not yet induced the Pigeon and the Guinea Fowl to consort with more than one partner; and the Swan, called Domestic by some writers, remains obstinately and even fiercely faithful in its attachments.

"The Domestic Gander is polygamous, but he is not an indiscriminate libertine: he will rarely couple with

A flock of Pilgrim breeders. Note difference in color between males and females. Government Experimental Farm, Ottawa, Canada.

females of any other species. Hybrid, common Geese are almost always produced by the union of a Wild Gander with a Domestic Goose, not vice versa."

The tendency to have more than one mate is more apt to appear in the lighter varieties, especially the Chinese, than in the heavy, more moderate moving Toulouse and Emdens.

Geese are sometimes difficult to mate. It is well to pen them together well before the breeding season. They seem to need to become thoroughly acquainted before deciding to mate. Apparently they consider it a serious undertaking; as though they realize it is for life and so they seem to give the matter deep consideration and make haste slowly.

The best way is to select the pairs or trios you wish to have breed together and isolate them at least six weeks before the breeding season; longer if practicable.

If they have been previously mated, make sure their former mates are out of sight—and sound. The goose or gander which has been mated before is loth to take unto himself a new mate with the old one looking on or within hearing distance. Apparently they think "it's well to be off with the old love before you are on with the new."

The safest way to ensure proper mating and good fertility is to make your selections in the Fall and keep the pairs or trios by themselves as far as possible, until the breeding season.

Fertility is of major importance in the breeding operations with any kind of livestock. When geese fail to produce fertile eggs it is an especially serious matter; it means the year's work is lost.

To those familiar with the promiscuous breeding habits of chickens and turkeys, it is hard to understand the fastidiousness of geese in this respect. Owners often mistake its cause and so are unable to bring about the right conditions for its cure.

If a mated trio shows only fifty percent fertility the owner, if unfamiliar with geese, is apt to think the gander is low in virility—that he lacks vigor—but this, usually, is not the case. A gander with vigor sufficient to mate with one goose or to produce any fertile eggs is usually fully competent to fertilize the eggs from two or three. If he is lacking in vigor there will probably be no eggs fertile.

Where only part of the eggs from a trio are fertile it will usually be found that the mating is faulty. The gander has probably mated with one of the geese and is ignoring the other; that all the fertile eggs come from one goose and not a half from each as might be supposed.

It is almost impossible to correct the trouble in time to do any good once the breeding season has started. Removing the goose with which he has mated will usually do more harm than good. The fertile bird usually upset by the change will cease to lay and so instead of a fifty per cent fertility for the two geese, there will be none, as the gander separated from his favorite before he mates with the rejected one at all will probably waste weeks in hunting and calling for his missing consort. By that time the laying season may be over.

If geese are secured from one source and the gander from another with the assumption that the mating will be universal, as in the case of cocks and hens, one is very apt to strike a snag and find that a mating has been made with but one, or in some instances two of the females, where matings are tried with three or four females and a male.

Do not assume a hundred percent mating until it is actually proven; take a little time and study your geese. If the females all look alike to you, band them with different colored bands so you can tell them apart and then a little intelligent observation will tell you whether the gander is spending more time with one than the others. Watch closely at feeding times. Rejected females are apt to be bullied and driven off from the feed trough. If the pen is harmonious and the gander shows equal devotion to each of his mates, the mating is probably universal; but it is best to be absolutely sure.

Test his impartiality by removing one female at a time and watching his actions. If he seems contented with the one or more geese remaining and shows no interest in the one removed, he is probably mated to one or more of those left with him. If on the contrary, he becomes uneasy and distressed and shows no interest in the remaining goose or geese, he is probably mated only to the one removed and should either be left with her for the season or if it is highly desirable that he be mated with one or more of those remaining, he should be left with them until they appear happy and contented together.

Try this test on each female in the pen in turn and by this method make sure. Do not expect fertile eggs from a goose turned in with a pair or trio already mated. If it is desired to introduce a new female into the already mated pen, remove the other females and leave the gander with the new goose until they appear happy and contented. This may take anywhere from two to three weeks but it is worth it: However, it is evident from the foregoing, that this must be done well before the time of egg production.

Geese are long lived. I, personally, owned a Toulouse goose who lived and bred at twenty-six years and was killed by an ox-cart-wheel crushing her head. I have known a gander to successfully mate at fourteen but it is safer not to depend on old ganders.

There are many accounts of geese living to an incredible age. Bement, in *The American Poulterers Companion* published in 1859, reports the following on the long life of geese.

"Of all the stock brought up on farms, the goose lives to the greatest age and is noted for its longevity. There are records of some attaining to a century or more. Twenty and thirty years are common periods through which its life may be traced; but more than twice the latter period has been well proved to have passed over its head, without the least diminution of its value for breeding.

"In 1824 there was a goose living in the possession of a Mr. Hewison (England) which was then upward of one hundred years old. It has always been in the constant possession of Mr. H's forefathers and himself and on quitting his farm he would not suffer it to be sold with the rest of the stock but made a present of it to the incoming tenant, that the venerable fowl might terminate its career on the spot where its useful and long life had been spent.

"There was also a goose on a farm in Scotland of the clearly ascertained age of eighty-one years, still healthy and vigorous; she was killed while sitting on eggs by a sow. It was supposed she still might have lived many years and her fecundity appeared to be permanent. Other geese have proved fertile at seventy years.

'A farmer near this place,' says Mr. Stevens, writing from Clanville House near Andover, England, 'tells me he has a goose now twenty years old and that she has hitherto never failed in hatching out two good broods every year; her second hatch this very season (1853) was ten, of which all are alive.' "

On this Dixon says: . . . "Many instances of the

longevity of the Goose are on record, and it is needless to repeat them. I have myself seen one upwards of thirty years of age, followed by a thriving family; but they are capable of reaching double and treble that extent of life. Indeed, the duration of the existence of the Goose seems to be indefinitely prolonged, and not terminable by the usual causes of decay and old age, (like Pliny's Eagle, which would live forever, did not the upper mandible become so excessively curved as to prevent eating, and cause death from starvation;) and reminding us of the accounts, apparently not fabulous, which we hear in these modern times, respecting the Pelican and the cartilaginous fishes. One thing is certain, that housewives do not consider Geese to be worth much for breeding purposes, till they are four or five years old. They will lay and produce some few young ones in the course of their second summer; but older birds fetch much higher prices as stock."

Geese breed well, as a rule, as long as they live and are active and vigorous but ganders are at their best from two to six years old.

I hold to my heart when the geese are flying
A wavering wedge on the high, bright blue
I tighten my lips to keep from crying:
"Beautiful birds, let me go with you."
          —*Grace Noll—Wild Geese, Stanza 1.*

CHAPTER VII

Hatching and Rearing

After the geese are mated about ten days, begin to save the eggs for hatching. Incubation is probably the most troublesome part of all the operations in the raising of geese. Sometimes the first eggs will be laid anywhere about the place and a careful watch should be kept that they may be brought in before they become chilled as geese usually begin to lay in late winter or early spring before the warm weather becomes settled and chilling will kill the fertility.

After the first few eggs, which may occasionally be deposited promiscuously, the goose will select a place for her nest and will deposit them there: in the majority of cases, she will choose her nest before she begins to lay. The eggs should be taken into the house or cellar each day and kept at a temperature of from fifty to sixty degrees if possible, and turned every day. They may be kept as long as three weeks if they are treated thus and turned regularly. Turn them half over each day; mark an O on one side and an X on the other and thus ensure turning them half way each day.

Each egg should be marked with the date laid, in order that they may be set in the same order to ensure against any becoming too old to hatch.

It is well to provide places suitable for nesting. Place as many barrels turned on their sides as there are laying geese. In these, place some hay or straw; the geese will each usually select one of them for a nest.

They will be more apt to select the nest beforehand and lay the eggs in the nest at the very beginning if

111

the nest places are provided well in advance, giving them an opportunity to become accustomed to them, some time before starting to lay.

If the geese are penned at night, low boxes with a little hay or straw within the shed or building in which they are housed, will serve nicely.

If the eggs are left in the nest, the sight of the gradually accumulating "clutch" seems to stimulate the mother instinct and brings on the sitting fever, so if the largest possible number of eggs is desired, they should be removed each day.

Goose eggs cannot be successfully hatched with turkey or chicken eggs in incubators, but there is no trouble hatching them artificially, if in incubators by themselves and proper temperature and humidity is maintained. The temperature should be about two degrees lower than for chickens or turkeys and the humidity much higher. If these precautious are taken, goose eggs may be hatched in incubators as successfully as those from chickens or turkeys.

However, for the average person with a small flock, the simplest and probably the most satisfactory way is to hatch the first eggs under hens and those laid later in the season right under the mother goose.

When the goose has chosen her nest she will proceed to line it with soft down plucked from the underside of her body and will cover the eggs completely every time she leaves to feed or drink. She needs no instruction in the art of nest building and the nest should be touched as little as possible during the laying period.

Let her sit right where she lays the eggs and she will take perfect care of them, leaving them each day for the proper cooling period, and will bring out the goslings in from 28 to 31 days, depending on the outside temperature, the fertility of the eggs and the breed of geese. The large Toulouse, Emdens and Africans as a rule require a longer period than the smaller, quicker maturing and more active Chinese and other smaller varieties.

If in dry weather, it will help, to sprinkle the eggs with warm water each day for a week before hatching; especially if the eggs are under hens. If under a goose and she has access to water every day, where she can wet her feathers when cooling the eggs, sprinkling the eggs is not necessary as nature will take care of the moisture problem in nature's way. If the eggs are removed each day from the nest as laid, the goose will sometimes become discouraged and seek a new nest. To prevent this, provide her with a nest egg and for this, nothing is better than a glass bottle. A beer bottle is about the right size and for some unknown reason, the geese seem to much prefer those that are green in color.

Let me repeat. Do not try to hatch goose eggs in incubators with hen or turkey eggs. Either the goose eggs will not hatch well from too high temperature and low moisture or the hen or turkey eggs will suffer from temperature and moisture suitable for the goose eggs.

Nine to fifteen eggs is the right number for a goose, depending on the size of the goose and four to six eggs for a hen. For the average size hen, five is right.

Rearing

Baby goslings are tender little creatures with a none too tenuous hold on life for the first few days; yet if given attention for just the first week or two, there is little trouble connected with raising them.

Goslings are virtually immune to disease. I have never known one to be sick from any of the diseases common to chicks or baby poults; but they must have attention the first few days to see that they do not get

separated from the mother goose or hen or away from
the brooder, if brooded artificially, as chilling may be
fatal to them.

Keep them out of water for the first couple of weeks.
At first, the down is dry and fluffy and water pene-
trates it rather easily and a chill is bad for goslings. If
a sudden rain comes up or if they are led through tall,
wet grass by mother hen, the soaking will do them no
good. Outside of the danger of chilling there is little to
worry about. Do not feed for 36 hours after hatching
—see that they then have sharp sand, water and tender
green grass if possible. Give then, a little moist mash
of any commercial starting brand or just cornmeal and
wheat bran, equal parts, will do quite well. Feed them
every two hours the first three days, gradually increase
the intervals until two weeks old when three times a
day is plenty.

The little fellows will nip the short, green grass the
day they first come out from their 36 hour rest after
hatching and after the first week this is all they need if
there is plenty of it, and if you only plan on raising
average "run-of-the-mill" geese.

An experiment on rearing geese for the Christmas
holiday on grass alone, and with supplementary feeding, published in the *Harper Adams Utility Journal,*
England in 1941, reads as follows: ". . . demonstrated
that it is possible to rear goslings . . . by allowing them
to graze, without any additional feeding. Results were
not such, however, as would be regarded as satisfactory
in normal times, when ample supplies of food are
available, and it is questionable whether earlier killing
and cold storage might not be a better procedure in
certain circumstances. The birds thus reared were yellow in flesh, stubby and not in such good condition as
the birds which had been trough fed. They showed a
very great improvement after being fed for 2 weeks,

however, and realized the same price per pound as the other birds.

"The birds that were given extra food for only the last 4 weeks were at the time of marketing practically the same weight (in fact they were a little heavier) than those which had been trough fed for 8 weeks. The experience with these groups was that the effect of feeding after the grazing period was an immediate and rapid increase in weight. The high rate of gain persisted for only a few weeks, however, and was not maintained with a continuation of feeding, probably on account of a diminution in the attraction originally aroused by the provision of trough food, and a consequent falling off in appetite.

"Finally, the over-riding influence of seasonal conditions for growth of grass cannot be too strongly emphasized. The extent to which geese can be reared successfully on grass alone depends to a great extent on the quanity and quality of herbage, which are governed largely by the weather and by skillful management.

"The experiment will be repeated, with some slight modifications, in the current year."

If you want extra size on your geese when grown, why that means extra feed while growing and the mash fed three times a day after the first two weeks should be continued twice a day, and all they will eat, in addition to plenty of tender, green grass right through the summer.

After the first three or four weeks they may be let into the water as often as they want and if no pond or stream is accessible, fresh, clean water should be supplied them at least twice a day.

Sun during the hot mid-summer weather is hard on the growing goslings and also on the old geese. Shade should be provided in some way. If there are no trees or shrubs on the range or pasture, set stakes and stretch burlap to make a canopy under which the geese may rest in the middle of the day, when the sun shines brightest and the temperature climbs too high.

The growth of goslings is exceedingly rapid. I know of no fowl which grows as fast. Specimens of the larger varieties sometimes reaching the amazing weight of four pounds in the first five weeks and up to eleven pounds in ten weeks. The smaller varieties do not put on weight quite as rapidly but they grow about as fast in proportion to their ultimate size.

Plenty of food is of course essential to rapid growth but bear in mind that unlike other poultry, a majority or all of this food may be grass, the most inexpensive of food. The ultimate in rapid growth involves, of course, grain feeding in addition to grass but satisfactory gains may and are made by successful goose raisers on grass alone.

The ideal range for growing goslings is a good pasture or meadow in which the grass keeps green all summer with a running brook for water. They will practically raise themselves after the first week or so, if left with the mother that hatched them.

If hatched with an incubator they should be brooded in brooders the same as chickens or poults for the first two or three weeks and then unless hatched very early, they require no artificial heat.

An orchard makes a good pasture. It provides shade during the hot weather and the geese are of great benefit to the trees.

My friend, John Reid of Massachusetts raises a sizable flock every year and pastures them in his apple orchards where they benefit the trees by eating the windfall apples, worms and larvae included, just as fast as they fall from the trees.

As a result, the insect pests grew less each year until

after a number of years of this practice Mr. Reid tells
me he does not spray for insect pests at all and the
trees are wonderfully thrifty in appearance, the geese
also rendering a service to the orchard by enriching the
soil by their droppings.

Geese mix perfectly with cattle in the pasture; but
do not make the mistake of filling the pasture with a
capacity lot of cattle and then adding the geese. There
is just so much grass and if there is but enough for the
cattle, it is quite manifest either geese or cattle will
suffer.

To the old saying that geese will make pasturing
unfit for cows, there is no truth. Obviously if geese are
closely confined on an area, completely covering the
ground with their droppings, the cows will not graze
there. Neither will geese if cows have fouled the
ground to the same extent. If there is plenty of range
and an abundance of grass, geese and cows will graze
together perfectly and both do well.

If grass is abundant, geese do not require high or
expensive fences. Thirty inch wire fencing is plenty
high for the heavier varieties or Chinese, which fly very
little and if the light varieties such as the Pilgrims, fly
over a thirty inch fence, the web may be stripped from
the flight feathers of one wing with a pair of scissors
which will effectually stop the trouble. A quicker way
is to simply clip the entire flight feathers from one
wing with a pair of shears but this shows and somewhat mars the symmetry of the body shape, while
stripping the web from the quill is hardly noticeable
unless the wing is extended.

It is only necessary to clip one wing whichever
method is used.

Housing

Housing geese is a very simple problem. Cold weather does not effect them. In fact, if they have shelter available, they will hardly ever use it except in the very worst storms in winter. However, they should have low sheds or buildings of some kind in which they can go in snow storms as their feet should be kept off the snow in cold weather.

If the ground is covered with snow and there is no shelter available, throw out some cheap hay or straw right on the snow, for them to sit on, and the geese will be contented and apparently, perfectly comfortable.

A breeding flock of American Buffs. Golden Egg Goose Farms, New Haven, Conn.

While housing is one of the very major problems with the production of eggs and chickens, it's a very simple and easily answered question in the raising of geese.

Be careful in handling geese. Never catch them by the legs as some do chickens. Their legs are light in

bone and rather tender and the heavy weight of the
body is apt to strain and lame the birds if held by the
legs. Always catch by the neck. Then hold the neck
with one hand and reach one arm around and support
the body with the other when carrying.

A wooden crook like a shepherd's crook, shaped
about like the chicken catchers so generally used, is the
handiest method of catching geese. Spot the one you
want in the flock, get them in a corner and reach out
and hook the bird around the neck and draw it toward
you when it may be caught with little trouble and
without alarming the rest of the flock too much.

The "goose-crook" may be made of wire but should
be wrapped with cloth so as not to hurt the bird's neck
when catching. A better way is to go into the woods
some day and cut a straight Hickory, White Oak or
Ash sapling about three quarters of an inch through at
the base, trim and cut off the tip, mark the shape of
the "crook" on a board with crayon, drive nails around
this pattern and then bend the small end of the sapling
around these nails and tie in this shape 'till dry; and
you have a lighter and handier implement and one
that will last a life time. It takes a little trouble but
it's worth it.

We charged upon a flock of geese
And put them all to flight.
            —*Young Soldiers, Stanza 3*—
            In McGuffey's Reader.

A Midwestern flock of Farm Geese.

CHAPTER VIII

Fattening

Fattening geese by natural method is a simple matter. When geese are raised by the economical, grazing system, the fattening process makes about the only feeding expense in raising.

After the first few days on a simple, moist mash for the baby goslings, grass is the only requirement for the growing birds until the fattening time in the Fall.

Let me again impress on the reader that if the goslings are to be raised on grass, there must be plenty of grass all Summer. If on pasture that dries up under the hot summer sun, green crops such as rape, swiss chard or other forage crops should be planned for and raised to take the place of the grass, the perfect food for geese.

Plan to have your geese ready for fattening about two to three weeks before the time you wish to market them. By this, I mean to hatch them so they will be nearly grown when the fattening time arrives. Growing geese do not fatten as quickly as the feed goes to growth instead.

By the "natural method" referred to in the opening paragraph, I refer to the common method in use where the goose eats just what it will and not the forced or "noodle" system of fattening which is in vogue in this country to a limited extent even now and which was the common way for geese to be fattened in European countries in earlier times; and to this day.

If the geese have been raised on the farm and have reached the age of four months, fattening is a simple and inexpensive matter. No extra buildings or pens of any kind need be provided. Whole corn makes an excellent fattening ration and if fed three times a day with plenty of clean, fresh water for the geese to drink, they will fatten satisfactorily and quickly.

If geese are brought for fattening, they should be "rested" in their new location for about a week and fed very lightly. Then gradually get them onto the fattening ration and finally give them all they will eat three times a day.

Corn, boiled, is softer and easier to digest and they will consume more of it with a consequent faster gain in weight. A mash of three parts corn meal and one part wheat middlings, moistened with skim milk is excellent or if skim milk is not available at a low price, water may be used.

The fastest possible gain will be made if the geese are penned in comfortable quarters with little opportunity for exercise and with all the mash they will eat three times daily and with fresh water before them constantly; or the corn, either boiled or not, may be fed in water and kept before the geese constantly, letting them eat at will.

Many thousands of geese are shipped into sections adjacent to New York City from the middle West each Fall and are fattened and trucked into the New York markets at a good profit to the farmers who carry on this business.

The method here is to pen the birds, about twenty each, in slat bottomed pens through which the droppings can fall to the floor beneath and to keep boiled corn in water before them constantly. They gain from two to three pounds during a three to four weeks fattening period.

The essentials in fattening geese are:

1. Have them the right age when starting to fatten.

2. Keep them quiet and contented.

3. Give them all the feed they will eat.

4. Keep fresh, clean water before them constantly.

5. Do not try to fatten them for over four weeks at the longest.

Time the fattening period to end when you wish to market your geese as after about three to four weeks they are apt to "go off" in appetite and consequently do not put on flesh economically and may even fail to gain in weight at all and thus no return is made for feed consumed.

The noodling system generally used in Europe to produce the tremendously over-fat geese for the enlarged livers used for pâté de foie gras, is as follows: Mr. Ernest E. Burel of Klamath, Calif., formerly head chef in noted hotels and a member of a famous Strasbourg family of manufacturers of pâté de foie gras, informs me that in 1937-38 more than 700,000 lbs. of this delicacy was exported from Czechoslovakia and Austria to Alsace alone at a price of $1.25 to $1.50 per pound. He says further: "Up to about forty years ago, this forced feeding for big livers was done by hand. To noodle geese, the farmer ground corn, added potatoes and cooked the mixture to a thick paste; this was rolled into noodles about two and a half inches long which were then dried.

"In feeding, the goose was held firmly and with one hand, the farmer would hold its mouth open, one finger holding the tongue of the animal in place, then the corn noodles would be inserted into its mouth and the farmer would massage them into its craw until well filled.

"This procedure was repeated three times daily. The goose was allowed to drink as much water as it wanted but it was confined to a little stall where it could not move about, having just sufficient room to stand upright.

"One man feeding geese by hand in this way could handle about six to eight geese per hour. . . . This method of noodling was quite generally discarded about 1900 (although it is still practiced in a few remote districts in Poland and Pomerania) and the following system, although short-lived, was established.

"The European goose feeder used a hand-operated sausage filling machine to do the work: The place where the butcher held his sausage-casing to fill the sausage into it, the mouth of the goose was inserted and a corn mash (of a similar consistency as the paste used for noodling) was fed. With this system the farmer was able to feed from 12 to 15 geese per hour.

"Then, in 1915, it was found that geese would produce larger and better livers if they were fed boiled, cracked corn and were allowed to drink all the water they wanted. A machine to feed the dry corn, operated by an electric motor was brought into use and is still in use today.

"This machine is a three legged affair having a cast iron container on top to hold the corn. On the lower end of the container is connected a transport-spiral enclosed in a brass housing; the latter rigidly attached to the machine or container. The transport-spiral is connected to the motor by a foot lever. Both the housing and the spiral are about 14 inches long, tapering from about 1¼ inches to ¾ inch at the mouth. The head of the goose is slipped over the brass housing which is placed deep in its throat and the spiral places the corn into the craw. A good operator can feed about thirty geese or more per hour by this improved method.

"A young goose fattened by this method weighs on the average from 15 to 17 lbs. without the liver. The liver weighs from one and a quarter to one and one half pounds. Livers of less weight are not considered first quality."

My friend, Harold Tesch noted breeder of Toulouse geese from Wisconsin, tells me hand noodling is practiced to this day in his state and that he considers it not cruel at all if properly done.

Mr. Burel states further: "Pâté de foie gras was first prepared by chefs of the royal households some 300 years ago, and although today the manufacture of this product has become a large industry, it is still prepared according to the formula followed by these chefs. It is a 'patty' made from goose liver combined with pork, flour, butter, spices, etc.

"Until 1900 most of the pâté de foie gras was manufactured in Strasbourg on the Rhine, only two firms having moved to Paris after the war of 1870, namely, the *Maison Henry* of Strasbourg and Paris and *Maison Olida* of Paris. Since then, however, some 60 to 70 factories have been established in the principal cities of Europe, and there are a number of smaller establishments in the provincial cities.

"In order to produce the large goose livers which are required for the manufacture of pâté de foie gras, young geese are either raised on the farm or purchased, and when they are almost full-grown, are confined in small cages to prevent them from moving about and working off nourishment. The geese are force-fed on a corn ration during the winters which are long and cold.

"Although the procedure of force-feeding geese for big livers has changed during the last 40 to 60 years from hand labor to the mechanical process, pâté de foie gras is still prepared in the same manner it was more than 300 years ago. However, a few new products have been added to the old standbys, namely, pâté and terrine.

"The average European pâté de foie gras manufacturing establishment uses from 200 to 250 pounds of goose liver per day in the manufacture of a variety of products, such as:

6 sizes of long pâté
10 sizes of round pâté
15 sizes of terrine
4 sizes of gibier and foie
8 sizes of fayennce
4 sizes of purre and pain
4 sizes of foie gras au naturel
4 sizes of parfait
2 sizes of supreme

and a number of fresh products which cannot be preserved permanently as Medaillon, Lucullus, Delice and others.

"An establishment purchasing 250 pounds of goose liver per day at the rate of about $1.25 per pound would also use about $50.00 worth of other products, such as pork, pork fat, butter, flour, truffles, containers, etc.

"Truffles are not an absolute necessity in the manufacturing process, but are included in certain high-priced products such as Parfait or Supreme, both being about the most expensive in the trade. Truffles are a fungus, found in the province of Periguex, France, and sell for about $3.50 per pound. In a pâté de foie gras selling for $5.00 to the trade, there is added about one-quarter to one-half ounce of truffles. An establishment using 250 pounds of goose liver per day would also use about seven to eight pounds of truffles. Truffles can be purchased either fresh or in cans. Many products produced in the pâté de foie gras industry contain the peelings of truffles or none at all. Truffles do not impart any decided or outstanding flavor to the product.

"Such an establishment as outlined above would employ 12 men and women to prepare the products ready for shipment: one chef, three helpers, one oven man, one dish washer, one can sealer, three women for labeling, two shipping clerks. The net sales value of the products prepared, depending on the daily variation of larger or smaller sizes produced, would be about $2,500.00. This figure does not include daily operating expenses as light, water, fuel, etc. The labeling of the finished products is still done by hand, and due to the different sizes of containers this procedure will probably not be changed.

"The building used by an establishment would include three large rooms with work-tables, one bakeshop, one butcher-shop, one packing room, one ice box, storage space for empty containers, one storage room for the manufactured products, and a sealing room which contains also the steam cooker.

"All in all, pâté de foie gras is not produced under great expense. It is a high-class product manufactured under an elementary formula. The chef in charge must have a time schedule for the cooking process, and this is one of the most important items. This time schedule is zealously guarded by the Strasbourg establishments. The various ideas about seasoning are merely a matter of taste. The goose livers and truffles, if the latter are desired, constitute the most expensive items, and they are used judiciously.

"Americans have been quick to imitate and produce certain products made for years exclusively in Europe, yet so far as the author is aware no American has ever attempted to pioneer in the pâté de foie gras field, where a high return could be achieved once an operating unit would be functioning smoothly.

"It should be as profitable to breed and raise geese to the perfection Americans have achieved in chickens and ducks, with the added advantage of having a high-class product to produce such as pâté de foie gras. America has everything in its favor to produce force-fed geese and goose-livers: land, corn, pork, pottery, cans, flour, butter, and a sufficient number of people to buy pâté de foie gras.

"With the inventive genius of the American people in overcoming high costs of production, it would seem that the large goose livers could be produced the same as any other product. A production line would not be an impossibility.

"To mechanize the force-feeding of geese, a 'ration adjustment' might be attached to the machine now being used. This would do away with the chance judgment of the individual operator. The supplying of corn to the machine could be done by overhead trolleys direct to the hoppers, and in general the feeding process could be mechanized in such a way that one goose would be fed per minute. Also, instead of buying young stock, the geese could be hatched, raised and fed on the same land and within a certain radius.

"For the disposal of the carcass after the liver has been removed, there are a number of channels open. A certain amount could be sold on the open market, the breasts could be salted and smoked and sold as 'smoked goose breast' (Pomeranian goose breast): others could be sold as boneless roasted goose in cans, etc.; the goose feathers or 'down' have a high market value also.

"The entire industry could be treated as two main units: (1) the breeding, raising and force-feeding of the geese, and (2) the manufacture of pâté de foie gras. Each unit would be operated on the books as a distinct revenue producing unit, the first unit selling its products to the second. The main stumbling block might probably be an oversupply of geese, but this would not occur if the producing of the carcass is taken into account from the outset. In Europe it is the Jewish population who takes these well fed geese during the winter months, there being a steady demand for the geese, and there is no objection to accepting the carcass with the livers removed.

"A goose-producing unit capable of breeding and raising about 2,000 geese per year could keep a sizable pâté de foie gras industry going during the season, from September to May, which are the months of actual production and manufacture of the pâté de foie gras. During the summer no manufacturing is done, but the products prepared during the winter months are then sold."

Thus far, we have no record of the manufacture in America of this delicacy and it looks as if, with our abundance of land for grazing, corn for fattening and high standard of living to furnish the demand for this product, that right here is a field for someone to pioneer the manufacture of pâté de foie gras into a profitable and useful business.

Egg Production

In the production of goose eggs there is also a chance for someone to do valuable constructive pioneer work. One of the most common questions asked about geese is "how many eggs will a goose lay?" It's just about as hard to answer as "how much does a house cost?" In fact, it's much more difficult.

A modern house costs somewhere in the neighborhood of a thousand dollars a room and if the questioner will specify the size house, one may give an approximate answer. Not so with the egg production of geese. There is no rule to go by. I had two good young American Buff females mated last year. One laid 14 eggs and one 34. Exactly the same blood and the same feed and conditions.

When we say a goose will lay from a dozen to sixty eggs according to age, it sounds pretty vague and indefinite but that's about as close as one can estimate. A little consistent selection for higher production during a short term of years would I am sure result in a greatly increased number of eggs and this would be an interesting and worthwhile piece of breeding work.

Ducks, which by nature are not heavy layers, have been by careful, selective breeding, improved to the point where in some of the British and Australian laying tests they have outlaid the hens and made records of much more than three hundred eggs in one year.

The possibility of greatly increasing egg production in geese by intelligent breeding work was brought forcibly to my attention by my experience with a young Pilgrim goose which Mr. Oscar Grow sent me a few years ago. Mr. Grow had been improving the laying qualities of his Pilgrims by selective breeding and wrote me when he shipped the young goose that she was from the best layer he had yet produced.

This young Pilgrim goose in her first year laid 126 eggs; a record as far as my knowledge goes for any breed of geese at any age.

The Pilgrims are not heavy layers as a rule and young geese lay less than in later years, so this phenomenal record aptly illustrates the possibilities in increased egg production in geese by intelligent selection of the breeding stock.

It would be a very simple matter. Hens have been developed to where high producers in breeding work have to be trapped practically the year round. In geese which lay only late winter and spring with an occasional clutch in the fall, they would require attention and records of eggs for only a part of the year for the first few years of pedigreeing at least; and if in small numbers, would really require no trapnesting as they will, as a rule, choose individual nests.

If trapping is found necessary, small yards or pens could be made and the geese identified by numbered bands on their legs and separated at night as the eggs are usually laid during the early morning hours.

Where the average production of a flock might be thirty eggs per season if this could be by a few years selective breeding, increased to fifty which is perfectly feasible, it would mean, at the minimum price for goose eggs for hatching of 25¢ each an increase in annual income of $5.00 per goose which would adequately feed the breeders for the year. There is no basic reason why geese that would lay 200 eggs a year should not be developed just as in hens and ducks which in the primitive state laid few if any more eggs than geese.

In this connection, let me speak of the possibility of marketing goose eggs as a regular food supply. Goose eggs are, in my opinion, much preferable to duck eggs as food. They have none of the duck egg flavor which, while some folks prefer it, is distasteful to others, nor do they have the tough, rubbery white characteristic of duck eggs. Goose eggs are much more delicate in flavor and aside from size cannot be distinguished from the finest hens eggs.

Only this morning I saw displayed in Carsons', a good restaurant in New Haven, Conn., goose eggs as a special feature. I asked the waiter the price and he said, "25¢ each, served any way" and added "they are three times as big as hens eggs you know." And this was an understatement. Eggs from well fed mature geese weigh from 7 to 9 ounces.

This job of increasing the egg production in geese is a good project for some of our agricultural colleges and would be an interesting demonstration in genetics as well as a service to the poultry industry.

CONSIDER THE GOOSE

"Yet well fare the gentle goose which bringeth to men even to his door, so many exceeding commodities. For the goose is man's comfort in peace, sleeping and waking. What prayse soever is given to shootinge, the goose may challenge the best part of it. Howe well doth she make a man fare at his table! Howe easilie doth she make a man lye in his beede! Howe fit even as her feathers be onlye for shootinge, so be her quills onlye for writinge."

—*Roger Asham, 16th century.*

CHAPTER IX

Exhibiting

Here is a field for a lot of missionary work by the folks who are interested in the goose.

I think there is no part of a Fair or Exhibition that attracts more interest on the part of John Q. Public and his family than the goose section. Still usually the geese are pushed off in some dark and obscure corner of the show room, crowded into the cages about right for Leghorn chickens and are looked on generally as a sort of necessary evil by the show management. There are of course exceptions to this general rule and such shows we should commend and patronize.

In the matter of judging, geese are usually assigned to some judge whose classes may happen to be light and so has some time on his hands. He may be qualified; or he may know little about geese, and care less, but the average show manager hardly takes this into consideration. The geese are there and have to be judged and the first available judge is handed the card.

Exhibitors should ascertain beforehand that suitable location, cooping, feed and especially water dishes will be given the geese and that a qualified judge be assigned to place the ribbons.

If such assurances are not forthcoming from the show management, keep the birds at home and exhibit at some show where this class is better appreciated and provided for.

Give them extra good care and feed generously for six weeks before the Show. If they are inclined to be wild, keep them penned for a week or two beforehand

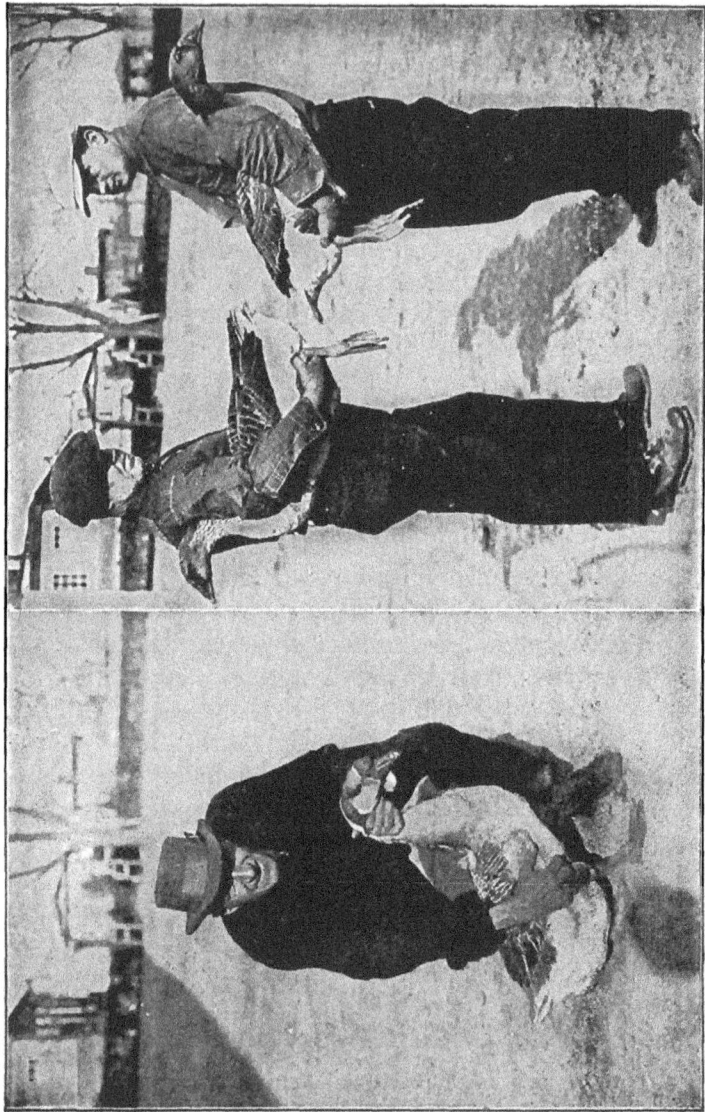

Proper manner of picking up and carrying geese with the head and neck under the arm. (Photographs from the Bureau of Animal Industry, U. S. Department of Agriculture.)

and move about among them quietly when feeding. They will soon become tame and docile and will show to much better advantage than if placed in the exhibition coops, wild and nervous.

The *American Standard of Perfection* sets forth specifically all the points desirable in domestic varieties of geese and should be carefully studied first of all when contemplating exhibiting geese.

Note carefully the different sections and do not fail to study and get into your mind the scale of points with its relative importance of the various sections in judging for perfection.

Note especially that shape, size, health and vigor account for 75 points of the total, or perfection, and only 25 for color in white varieties and that 63 for the essential points named above and 37 for color in the breeds other than white.

Good judges of geese will place heavy emphasis on the important economic sections of health, vigor, size and shape and will let the less important sections of color of plumage turn the scale only when other points are approximately equal in two specimens.

Geese are raised primarily for food and so it is important that your geese entered for the show be up to weight. In the big varieties of geese size over Standard weight is important only if it is well balanced and symmetrical. Mere weight caused by over fattening should never be considered a desirable quality.

In the smaller breeds such as Sebastopols, Pilgrims, Chinese, the birds nearest to the Standard weights should have the preference as they are bred to fill the demand for smaller sized geese and the weights given in the Standard are the ones found by breeders with long experience to be the most suitable.

Make sure your geese are in good flesh and in good feather. Above all see that they come into the show room clean. Exhibitors go to great ends to wash and fit chickens for showing but none of this washing by hand is necessary for geese.

Give them plenty of time before the show and plenty of water and they will do the job themselves. They do not require a large pool; just furnish them a pail of clean water two or three times a day for a month before the show and place this on the grass or some place that will not become muddy and they will clean themselves and be in fine shape plumage-wise, when the show opens.

Have your geese fat but not over-fat when you show them. All stock looks better when in good flesh and geese are no exception to the rule.

Use proper shipping coops in sending to the shows. Showing is a publicity measure primarily and nicely painted shipping coops with your name and address plainly lettered on the sides will attract favorable attention enroute as well as when they arrive at the show and the more attention your geese get, the more publicity for you and your business and it all helps sales.

Have the shipping coops plenty high. A coop for a pair should be at least three feet high and four feet long and not over 14 inches wide. This gives them plenty of room to stand erect and does not permit them to turn around and break their tail feathers and ruffle the plumage generally.

If the trip to the show is short, do not plan feed or water enroute. See that they have a full feed and follow that with an opportunity to drink plentifully just before they start and they will arrive in good condition at the show and without the opportunity of wetting their plumage and the litter in the shipping coop on the way. But if they are to be more than six hours on the way, make provision for feeding and watering and post a card placed conspicuously on the shipping coop calling attention to the fact that THESE GEESE MUST BE FED AND WATERED. Make provision for this by having a good big cup or can attached to the shipping coop, either inside or out; preferably on the outside for the water, and tie on a small muslin sack of corn where the express agent will not fail to see it. A quart size tomato can is as good as anything for the water recepticle and is best placed on the outside of the coop with provision made for the goose to put its head through to drink. If placed on the inside, put it in the corner tacked to the sides and two feet or more from the bottom. If pairs of geese are shipped together a can should be placed in each of two corners. Another can or small box may be placed in the other corner for feed enroute.

Put plenty of good, dry material in the bottom of coops. Nice bright, clean straw, plane shavings or even sawdust will do; but have it at least three inches deep to absorb any moisture from spilled water or droppings so the geese will reach the show in clean and sightly condition.

Then when you coop them, see that they have plenty of litter in the exhibition coops and arrange with the attendants to have litter changed each day and see that the water vessels are adequate. How many times have I seen big geese in regular chicken exhibition coops not high enough to allow the birds to stand erect or turn around and with water cups about right for a bantam and usually as dry as a bone!

Too many shows employ attendants that fail to realize that a goose cooped in a hot show room will require as much water as five or ten bantams, but will use just the same sized water cups and furnish the same amount of feed and water to both.

See to it that they have some green food each day

during the show. Cabbage or root crops of any kind, cut up will do and the geese will keep in much better condition than when fed grain alone.

Have some attractive cards printed to display on your cages at the show; give the name of the breed and your own name and address and in this way get every bit of advertising possible from your exhibit.

It is expensive to show geese. The shipping coops have to be built; entry fees paid, shipping costs are about ten times the amount paid for bantams on account of the extra weight and the premiums are usually the same, so it is quite essential that you derive all the benefit possible from your investment in exhibiting your geese.

If possible, make a deal with the show manager before hand that the geese shall be shown in turkey size coops as the ordinary exhibition coop is much too small.

Finally, when the birds come home after the show, if in bitter, cold weather, keep them inside for a day or two until they become accustomed to the great change from the temperature of the show-room—usually much too hot for outdoor birds—to the cold winter temperature of the home farm.

CHAPTER X

Importance of Feathers

The custom of plucking live geese, in old days a universal custom, is not as prevalent now. It is still practiced in foreign countries and to some extent in sections of the United States where many geese are raised.

It is not necessarily a cruel process. Plucking the soft feathers from the breast, sides, belly and back may be successfully and humanely done in Spring, Mid-Summer or early Fall if care is taken to leave the under-down and especially the feathers under the wing which are required for their support.

About one pound may reasonably be expected from each goose by three pluckings, on an average and if a large flock is kept, this runs into a considerable sum as the feathers sell at good prices.

The feathers are ripe for plucking when the quills are dry and contain no blood. A little practice will teach just when they are right for the plucking operation.

A thin stocking should be drawn over the head of the bird which is held in operators lap while plucking.

Goose feathers are used in the manufacture of the finest mattresses and pillows and there are dealers in the larger cities who buy them.

There is no difference in the softness nor resiliency of the different colored feathers but the white feathers are preferred by dealers generally who will usually pay a higher price for them.

When the feathers are plucked put them in muslin

bags and hang them in a dry place where they will keep perfectly until sold.

Especially during times of war in European countries have goose feathers brought high prices in this country as great numbers of them are imported from Europe during normal times.

Never attempt to pluck geese unless the feathers are "ripe." This may be told by examining closely. When they begin to shed feathers and on examination, pin feathers are discovered, then is the time to pluck. The feathers are ready to come out and the goose suffers not at all from the operation. Moulting is a long process and plucking may be practised two or three times during the Summer and practically all the feathers thus saved which would otherwise go to waste. Never pluck your young geese and it is wise to pluck yearlings but once the first season.

The composition of feathers is 73.15 protein—21.6 water—1.01 fat and 1.01 ash.

By the old time French peasants, geese were kept for feathers mainly. The French people mix the goose feathers with hair and wool which they use in the manufacture of yarns and cloth. Many of these are used on the farms but many also come into good use commercially for stuffing pillows, cushions, furniture, quilts and mattresses.

The establishment of many military and civil hospitals is making strong demand now for goose feathers for pillows and comfortables. Cantonments and emergency quarters for soldiers and civilians in defense industries also add to the big demand for feathers. While all these needs do not require goose feathers, they are required for the very best, softest and lightest needs.

The goose is probably the only fowl ever domesticated and kept mainly for its feathers as it was for centuries by the French peasants.

Cleaning Feathers

In an article in the *American Egg and Poultry Review*, Herbert Bunyea of the United States Department of Agriculture gave the following description of rinsing, soaking and bleaching feathers—"As a routine procedure all feathers are soaked in a lukewarm soap solution to which ammonia, carbonate of soda or borax are added. This process removes ordinary dirt and frees the feathers from the grease always present in newly plucked feathers. After the cleansing process, the feathers are thoroughly rinsed and then soaked in warm water for thirty or forty minutes.

"Following the process of cleansing, feathers, white or colored, are frequently subjected to a bleaching process. This is done by one of several methods. They may be soaked for half an hour in a solution made by dissolving 20 grams of potassium permanganate in 50 liters of soft water after which they are drained on wood-slatted trays or trays of rust-proof wire-cloth; then rinsed in a cold bath of 50 liters of water to which 200 grams of sodium bisolphite and 30 cubic centimeters of sulphurous acid has been added. This treatment removes the brown residue left by the permanganate of potash.

"Optionally the bleaching may be done by immersing the feathers in a solution of hydrogen peroxide; 1 pint to six quarts of water slightly alkalinized with ammonia, gradually bringing the temperature of the bath to 120 degree F. However, it is said that peroxide bleaches may seriously damage the texture of the feathers. White feathers for coloring are not usually bleached beforehand; it is sufficient to scour them for the removal of blood, dirt and grease . . ."

According to the census of 1939 we imported 6,319,-659 pounds of crude feathers—value $2,251,676. which went into pillows, cushions, sleeping bags, aviators suits, upholstery, etc. Hungary sent the largest amount —473,420 pounds, value $345,049. at pre-war prices. France came next and a close second.

In case your feathers are not marketed promptly and become infested with insects the following advice by Prof. L. E. Weaver of Cornell University in the *American Agriculturist* will be valuable:

"The insects that are damaging your goose or duck feathers are probably either common clothes moths or carpet beetles. If the larvae (worms) are fuzzy they are carpet beetles; if they are in a smooth case they are moth larvae. In either case treatment to get rid of them is the same.

"You have a choice of two treatments.

Carbon bisulfide. Get this liquid at the drug store. Put the feathers in a tight container such as a 50 gallon drum. Put three tablespoonfuls of carbon bisulfide on the feathers and cover tightly. The liquid evaporates and forms a poisonous gas which destroys the insects within a half hour. This fumigation should be done where there is ample ventilation. Keep the feathers in a moth-proof container to prevent reinfestation.

"*DDT treatment.* For two bushels of feathers, use three ounces of 10% DDT powder. Dust the powder over the feathers in a large paper bag. A larger amount will do no harm, but the feathers will be dusty. This treatment protects against reinfestation for a long time."

CHAPTER XI

Marketing

The marketing of geese in this country is in just about as primitive and unorganized condition as their breeding and production.

There appears to be no general plan, but each section and even each goose raiser seems to dispose of its, or his, product as may be deemed most advantageous to supply whatever particular local demand there may be.

When sold for shipment in the middle west where most of the geese are produced, the raiser seems to be at the mercy of the local "henman" or "gooseman" who travels from farm to farm and picks up what is offered for sale at his own price and sells to the shipper at a central point who in his turn ships to the buyer in the large city and he to the retailer and finally the consumer gets his goose at long last and at a price which shows a profit for all concerned except the raiser and he usually figures whatever he gets is pretty nearly clear profit as the geese probably were raised on grass with a little corn to "finish off."

Occasionally there is someone who takes his geese more seriously like Mr. & Mrs. John Reid of Massachusetts, who raise a hundred or more a year and sell them all to a waiting list of private customers who know and appreciate the fine condition, perfect dressing and honest price of the Reid geese and wait for them each year. The Reids get a price somewhat below turkey but when the comparative cost of raising is considered, a price that is much more profitable for them. When ready for the market his customers are called by telephone and the orders taken, filled and delivered.

In New Jersey, many thousands are shipped in by freight from the middle west and crate fattened by farmers for about two or three weeks and then trucked to New York City to supply the very strong demand from the Jewish race for live, fat geese to be Kosher slaughtered according to the rules of the ancient Hebrew religion.

This business evidently has proven profitable as it has grown during the past twenty years from a small beginning to a big business, handling an immense number each fall.

I have never heard of any Governmental, State or Industry regulations or supervision in the matter of marketing geese and I have never known of any promotional effort put forth to increase the use of geese or goose eggs as food. Off hand, I can think of no other food product in which such efforts have not been made. Eggs and poultry, milk and cream, beef and pork, ducks and turkeys, cranberries and maple sugar all have been publicized and made attractive by improved marketing methods but the goose, in the matter of marketing is just about as far behind other farm products and animals as it is in the attention of educators and producers, in improved production and marketing methods.

A. C. Dingwall, Associate editor of *American Egg and Poultry Review* of New York City, in reply to a letter on marketing of goose eggs and geese in New York City writes under date of June 15, 1942: "Records on goose eggs and geese prices are not available due to the fact that they are so seasonable and for the same reason it is not possible to compare goose prices with other classes of poultry . . . very few goose eggs come

to the New York Market and what are received here
arrive just prior to Easter. Practically none are received at any other time during the year. In the past
season, prices for goose eggs varied widely depending
on size and quality. They sold wholesale from $1.50
to $3.00 per dozen. Retailers sold them from 15 to
30¢ each."

Dr. Alfred Van Wagenen in a study made at Cornell
University during the pre-war period, and published
in Bulletin 808 from that institution, finds as follows:

"The average size of goose eggs is about three times
that of hen's eggs. They are available on the market
only for a short period around the spring holiday season. Prices are extremely high just before the holiday
and drop to 40-60 percent of the pre-holiday price during the two weeks following it. Prices average 5 to 8
times those for best large white hen's eggs before the
holiday and about 2 to 3 times the hen egg price immediately following the holiday. As with duck eggs,
an early Easter results in higher price levels for goose
eggs, but the severity of the drop in prices after the
holiday is about the same regardless of the time of the
holiday.

PRICES FOR GOOSE EGGS FOR TWO WEEKS BEFORE AND
AFTER EASTER IN NEW YORK CITY

Two weeks before Easter	Two weeks after Easter	Per cent post-holiday price is of pre-holiday price
Cents per dozen		Per cent
90.6	45.9	51
147.5x	85.9x	58
188.0x	122.3x	60
179.9	77.0	43
229.5x	93.0x	41

x Comparable quotations for only 4 of 5 years available.

"Now as to geese; through the year, there is a wide range in price, supply and demand. In the course of a year, about forty cars of live geese are bought by the fatteners (about 1200 to a car) during the months of October, November and December. The receipts average about three cars a week. In the remaining months receipts probably average one half car a month. During the year prices vary greatly; 15¢ to 28¢ per pound. Demand and prices are best in October, November and December.

"Last year in those months, fat geese sold from 23 to 28¢ per pound. At the present time (June) prices for live geese are low; 13 to 15 cents . . . in the fall months I would say on all grades of geese prices would average 18 to 22 cents."

So, you see even in the largest and best organized marketing city in the country there is little, if any, system in the marketing of geese or goose eggs. No records of prices are kept and Mother Nature, supply and demand, and the distributors rule the roost as far as marketing geese and goose eggs is concerned.

In the middle West we find things just about the same.

G. D. McClaskey of the Kansas Poultry Institute has this to say on the subject under date of July 2, 1942.

"Geese are marketed wholesale both live and dressed. However, on account of the high price of feathers, they are mostly marketed dressed by poultry packing firms because the poultry packer finds that he is able to gain more revenue from the feathers than from the actual processing of the goose itself.

"Goose feathers are selling at the present time (Summer, 1942) for $1.50 to $1.75 per pound, which means they are worth considerable money.

"At one midwest poultry packing plant nearly $50,-

000 worth of goose feathers were sold last year (1941). 'So, when I think of marketing geese,' this poultry packer said, 'I think more of marketing the feathers than the geese themselves. For that reason I imagine other packers are more anxious to handle dressed geese than live ones. We are.'

"As a general proposition, each poultry packing plant handles some geese, but mostly without making any attempt to specialize in the handling of this product. However, when a poultry packing firm will segregate live geese at one feeding station, as is being done in some instances, where the geese are fed cooked whole corn in boiled milk, thus getting uniform results, it is much easier and far more satisfactory to handle them than it is to dress small lots in an amateurish way at several plants. This is a practical point that cannot be overlooked.

"Where the dressing and packing of geese is as definite a part of the business at a plant as is the dressing and packing of other poultry, the dressed geese are box-packed, from four to six birds to the box, depending on the size of the birds. The geese are frozen after being packed, are shipped in refrigerated cars, and are held under refrigeration until they go to consumers.

"One of the leading poultry packing firms in the handling of geese says the smaller goose that weighs about 10 pounds, dressed, is by far the most favorable from a chain store sales point. In recent years, he said, geese have become so large that some of the heaviest, running 16 pounds or more, have to be sold at a discount.

"Eviscerated geese have not sold as readily as other eviscerated poultry. One reason given is that it is presumed many buyers of geese probably want to keep the goose grease.

"The largest quantity of geese is produced in the

Middle West and, for that reason, their sales 'fan' the entire United States. In 1941 one midwest poultry packer shipped carloads of dressed geese to San Francisco and Los Angeles and, at the same time, was making carload shipments to Chicago, Boston, Detroit and New York City. This packer said that this wide distribution of dressed geese from the Middle West is not true in the case of distribution of any other poultry.

"Ten years ago geese were sold as a substitute for turkeys. The high price of turkeys at that time meant that geese, because of their lower price, could serve as a substitute, particularly among the foreign-speaking population. This situation has been changed in recent years because of the relatively lower price of turkeys. The goose market still depends considerably on the price of turkeys, but the differential is much smaller.

"Goose eggs always sell for more money than hen's eggs during the week before Easter, with the price perhaps being higher in New York City than elsewhere. At any other time, goose eggs probably move along with other eggs to the egg freezing and drying plants where they are worth more money, because of their yield, than they would bring on an open market basis."

A possibility of further income from geese and one that, if the general use means anything might well grow into a tremendous demand, is that of the skins of geese for milady's powder puff. Below is an article from *The Poultry Farmer* and *Feathered World* of Great Britain which gives food for thought. I quote:

"The world is a buyer of British powder puffs and powder puffs are made from goose skins—not swansdown, which is too soft.

"But the goose skins are not obtained from ordinary geese, nor, in fact, from any goose produced in England.

"The feather merchant who processes the skin in

this country wants a pure white, thick-downed skin from a 20 lb. goose. Moreover, he wants the skin, as received from the foreign agent, so free from grease that the down can be blown quite clear right down to the skin.

"This 'degreasing' is a skilled and secret process of the Continental merchant handed down from father to son, and chemists have never been able to achieve the same results, producing woolly and slightly greasy substitutes.

"This specialty goose production is confined to a certain district in France and also of Poland, where the right type of goose—in flocks of 1,000—and the correct method of processing are available.

"The foreign agent buying the skins needs considerable capital because all skins are bought between September and December (at a price pre-war of 3s to 4s).

"The question whether the same goose could be produced over here has attracted—and broken—many pioneers. It is not merely a question of the type of goose. It might be possible to increase the average English farmyard goose from 10-11 lb. to 20 lb. by using the Emden, and climate and soil can hardly have a pronounced influence on the thickness of the down. Nor can the degreasing process be an insurmountable obstacle.

"The real hurdle is quantity. Who is going to raise 1,000 geese? . . ."

When a boy at home, we used to sell all our limited output of geese to customers in the neighboring city and there was always a demand as long as they lasted. They were sold "full dressed" and ready for the oven nicely finished and neatly wrapped. The demand started with cool weather and lasted till well into the Spring.

Picking is a tedious job. The method followed by

my mother and handed down to the succeeding generation was to heat a wash boiler of water to the boiling point and immediately after killing plunge the goose in for about two or three minutes. Long enough for the feathers to get partially soaked and very hot. Then take the bird out and wrap in a woolen horse blanket and let it steam for five minutes or more. After this is done, the feathers will come out comparatively easy and in a quarter the time it would take to dry-pick. And the carcass will look plumper and much more attractive.

Plucking can be facilitated very much by use of the modern mechanical plucker. If dressing geese on a large scale, by all means secure one, but if only a few geese are to be dressed, it will not pay as these machines are high in price.

In most cities there are live poultry markets using these modern machines and among them are probably those that will do your plucking for a reasonable fee. In my own city I have found it so.

Geese should be first scalded just as chickens are, before being held against the cylinder of the plucking machine, but it will take some longer to saturate the thick down of the goose. To determine when adequately scalded, pluck a few feathers from the thigh.

Killing is simple; either by the sticking method as described for the killing of ducks in the latter section of this book, or by cutting the throat with a sharp knife; the more common method in the live poultry markets.

That a demand, and at a premium price, might be created for geese, is indicated by the Baltimore market where every fall the locally famous Kent Island geese are in good demand at a price several cents over the regular price of geese on the same market. These geese are grown by the farmers of Kent Island in

Chesapeake Bay, are raised on grass and nicely fattened in the fall. They are white in color, fairly uniform in size and because they are alike in color and size and always in fine condition, have built up a strong, preferential consumer demand. The famous "Mongrel Geese" of Rhode Island always brought a premium on the Boston Market of five to ten cents above the market price simply because they were invariably young, well fattened and marketed in beautiful condition by folks who made goose raising a business. They were the result of a cross between the grey geese of the Rhode Island farms and the Wild Canada geese. They were hybrids or "mules" and so were never kept over for breeding and thus were always marketed young and so their reputation was never damaged by old geese being included when sold. The same conditions would result in increased demand for geese in other places and shows the possibilities for advancement of the goose as a table bird.

Just as in other livestock and other food products of all kinds, a uniform quality, by good grading at the farm, with attention to proper feeding, growth, fattening and dressing would immensely stimulate the demand for this useful bird.

The last time I bought a goose on the market for Christmas dinner, it was almost unfit for food. It was tough and greasy and old. It smelled and tasted as if it had been in storage for some years following a long, useful life as a breeder and altho my family were considerate enough not to make too many remarks, no one ventured more than a taste and all were rather insistant that the next year we should repeat the Thanksgiving turkey at Christmas time instead of the traditional Christmas goose; and yet at a later time when we had a home-raised, fattened, dressed and served, young Emden, every member of the family helped to pick the bones.

The best way to increase demand for table geese is to be careful never to sell one that is not well fattened; but not over fat, young and nicely dressed. Old geese are never the most palatable and over-fattened geese do not appeal to most tastes.

Among the Jewish race, there is always a strong demand for very fat geese as the grease is used and highly valued for cooking but ordinarily over-fat geese are not in demand.

The economical and practical method of raising geese for satisfactory marketing is to give them plenty of good, green-grass range with an abundance of water and then when cool weather comes, fatten them up quickly with whole corn or wet mash and market them before they become over-fat or go off their feed. Do not start to fatten them until two or three weeks, at the most, before you wish to market them.

Establish a reputation for your geese by having them uniform in color, in size and always of the same high quality and there will be no trouble marketing them at a premium over the common market price.

Perhaps the fact that in recent years too much stress has been placed on size in geese may have held them back as a popular fowl for the table. The day of the big family in the better homes is past. Even the Government has been working for a number of years to develop a small turkey suitable for the small oven and the small family.

Joseph P. Quinn of the United States Department of Agriculture, has this to say on the subject, in a recent article in *Cackle and Crow*.

"Dr. D. C. Gordon of the Bureau of Animal Industry states that he believes that a small goose of superior economic qualities would be much more popular on the farm and in the holiday menu of the modern housewife." "This view is shared by other authorities" continues Mr. Quinn. "The late John Robinson" says he "has pointed out that it is contrary to the general principles of American poultry standard making to establish weights that exceed requirements for utility purposes; he believed that the mammoth show birds found in two of the most popular breeds of geese did not represent the most useful type nor the most desirable size of the modern goose. Weights described in excess of market requirements are described by him as being detrimental to the popularity of the breeds in question as well as to the extension of goose raising generally along profitable lines."

"According to Geo. D. Quigley of the University of Maryland and informal survey among Baltimore marketmen revealed a pronounced preference for a smaller goose."

"Kent Island geese quoted 22 to 24 cents per pound on the Baltimore market are considerably smaller than the Toulouse and are pure white in color. These geese are raised very economically on fresh and salt water marshes of the Eastern Shore of Maryland."

From the above and in keeping with the trend of the times it would seem that there is a distinct place for the smaller geese for family use and the larger varieties for hotel and restaurant trade just as in the private, hotel, restaurant or retail marketing of turkeys.

By finding the potential market; and producing the kind of geese best suited to it and presenting it in the finest condition and most attractive form, there is plenty of opportunity to develop a profitable market for geese in this country.

> Let the long contention cease!
> Geese are swans, and swans are geese.
> —*The Last Word, Stanza 2—Matthew Arnold.*

CHAPTER XII

Geese Must Have Grass

Here's the evidence

"Geese are fed in wet places and it is the practice to *sow, especially for their food supply,* using for this any kind of grain, but particularly that salad plant called endive which keeps green wherever there is water, freshening at the mere contact of water however dry it may be." *Varro's Rerum Rusticarum Libri Tres.* Book II *The Husbandry of Livestock.* 67 B.C. Translated by Fairfax Harrison in *Roman Farm Management.*

"*Let marshy land* also, *but such as is grassy* be destinated for them: and let all sorts of food be *sown for them* such as tares, trefoil fenugreek, but especially a kind of endive which the Greeks call ferin; the seeds of lettuce also ought, above all things, be sown for this use because it is an extremely soft and tender pot-herb and very much sought after and much liked by these fowls; as also very useful food for their young ones." Collumella. The Roman writer on agriculture who lived in the first century A.D.

"They (geese) are very profitable in places *where there are commons to feed them* on, being creatures that require little care and attendance and little charge in feeding them . . ." Willoughby's *Systema Agriculture,* London 1675.

"Some may fancy that it is less worthwhile to be very careful about this fowl (the goose), because their profit is less than that of the hen or the turkey; but the trouble required about them is also less and their profit is not so small as some affect to make it . . ."

". . . Besides water, a great advantage for the breeding is a *good quantity of common;* they will on these places provide for themselves in a manner, without the regard or trouble of the owner . . ." Thomas Hale in *Compleat Body of Husbandry,* England, 1756.

". . . in those parts where the cultivation of maize is held in high esteem and *where there are pastures,* the goose is what it was a century ago . . ." Main, *Treatise on Poultry* (In France) 1819.

"They (Geese) are scarcely profitable to keep *unless they have the range of an extensive pasture or common,* when they require no other kind of food." Micajah R. Cock, *The American Poultry Book.* Harper and Bros. New York, 1843.

"Where people have a right of *common* or live in the vicinity of *marshy heaths* the breeding or rearing of geese will prove very profitable or in such situations they are kept at slight expense." Dr. John Bennett, *The Poultry Book,* Boston 1851.

"The chief requisites for goose-keeping are a pool of water and a pasture for grazing. The latter is essential as the bird is graminivorous as well as graniverous." C. N. Bement in *American Poulterers Companion,* New York, 1856.

"With respect to the range and domestic accommodations of geese, they require . . . a green pasture or common . . ." D. J. Brown, *The American Poultry Yard,* Boston, 1859.

"Grass being an indispensible part of the food of goslings, we were necessitated to procure for them pieces of turf." Mr. Trotter, "The Journal of the Royal Agricultural Society" from *The Poultry Book*, W. B. Tegetmier, London, 1867.

"Grass seems to be their (geese) natural food and by following nature in all cases with animals and more especially with fowls, we have succeeded best." W. M. Lewis in *The Peoples Practical Poultry Book*, New York 1871.

"Grass is essential to the well-keeping of Geese." George P. Burnham, *Burnham's New Poultry Book*, Boston, 1877.

"(Geese) in a small way are most profitable . . . the stock birds and young ones during a great part of their growth cost scarcely anything as they *graze and forage* about especially on wheat stubble." Lewis Wright in *The Practical Poultry Keeper*, London 1901.

"They (Geese) are *essentially grass eaters* and will find practically the whole of their substance. The goose raiser should bear in mind that his business is not to substitute artificial for natural food but to give the former to make up for the deficiency in the latter." Sir Edward Brown, *Poultry Husbandry*, London 1915.

"*Suitable pasture* is the first requirement in profitable goose growing." John H. Robinson, *Growing Ducks and Geese for Profit and Pleasure* 1924.

"They (geese) are more easily and profitably grown than ducks, owing to the fact that they *subsist largely on grass . . ."* Louis M. Hurd, *Practical Poultry Farming,* New York, 1930.

". . . In fact they (geese) are classed as grazing fowls living chiefly upon grass and herbage about their haunts . . . Many varieties of noxious weeds are especially relished by them and for this reason they are in much demand upon the cotton plantations of the south; while in the cornbelt farmers are discovering that the geese may be advantageously turned into the cornfields after the corn has reached sufficient height; here they not only destroy many of the weeds but keep down the so called 'suckers' from the base of the hills . . . Breeding geese should be put out upon grass pasturage since *green grass is vital* to both goose production and high fertility . . . *Goslings must have access to young tender grass* from the beginning if they are to thrive well . . . and when this is young and tender they require very little other feed . . ." Oscar Grow, Famous Goose breeder from an unpublished book on ducks and geese.

"Geese can be raised successfully and profitably in all parts of the United States, but are most abundant in the middle and north Central States. They subsist very largely on grass during the growing season and are the closest of grazers; therefore they are most economically raised where pastures are abundant and where the grass remains green and tender during long seasons.—*Bulletin*—United States Department of Agriculture, 1933.

Suitable Forage Crops

Climatic and soil variations due to geographical locations makes it impractical to give pasturage directions that will be satisfactory in all sections of the country. For the northeast section of the country, Prof. J. R. Smyth, Head of the Poultry Dept. at the University of Maine writes as follows:

"I have consulted our Extension Agronomist and he makes the following suggestions: He would not consider either rape or soy beans as satisfactory crops for Maine. He suggests rye seeded not later than September 1st for early spring,—this to be followed by oats seeded as early in the spring as possible, probably early May. For fall, he suggests Jap millet rather than sudan grass for all of the northern part of the state and any place else where the soil tends to be wet. This should be seeded by the middle of June and will make good forage by August.

"This is going to leave some gaps and he suggests that there is nothing better than natural grass or ladino clover. The ladino clover should be seeded in the spring with oats but will make very little pasture that year. In fact, he suggests that good ladino clover will come the nearest to making pasture throughout the year of any crop we can grow in this state."

Some suggestions for Connecticut poultry pastures by B. A. Brown, Agronomist, University of Connecticut.

"*Soil Treatments*. Practically all of Connecticut soils are acid and, excepting on the well-fertilized tobacco, potato and vegetable farms, are deficient in readily available phosphoric acid and potash. Before seeding clovers and grasses for poultry ranges, fields not limed or fertilized within five years should receive one of the following treatments (amounts are for one acre):

(1) Limestone	2000 pounds
Superphosphate (20%) ..	400 pounds
Muriate of potash	100 pounds

or

(2) Limestone .... 2000 pounds
    Poultry manure ....   4 tons

Because limestone, and especially superphosate, penetrate very slowly into the soil, little loss from leach-
ing occurs and further additions of those materials
should not be made for, at least, five years. Properly
managed poultry ranges should require little additional
fertilizer, but a ton of limestone once in eight to ten
years would benefit most pasture species. Poultry droppings are relatively low in potash, which clovers need in
large amounts, and therefore the application of muri-
ate of potash at 100 pounds per acre each year will
help lengthen the life of the clover.

Seeding Ranges. Unless a good stand of clovers or
grasses is present, it is advisable to plow, harrow and
reseed. Grasses will usually make stands seeded as late
as September, but clovers may not live through the
winter if seeded after August 15. If unable to plant
until September, the grasses may be seeded then and
the clovers sown the following March. A firm, smooth
seed-bed should be prepared and the seed covered with
only about one-half inch of soil.

Grasses and legumes on poultry ranges should be
able to withstand close and frequent grazing for many
years. This qualification eliminates such good "hay"
plants as alfalfa, red and alsike clovers and timothy.
A seed mixture which may be expected to give satisfaction on limed, fertilized soil is given here (amounts
are for one acre):

Redtop 2   pounds
Kentucky blue grass .... 12-20 pounds
Ladino clover .... 1-2 pounds

Bluegrass and native white clover will usually vol-
unteer and eventually make a good pasture on run-
down meadows and pastures if lime and fertilizers are
spread on the surface.

Management.—Several factors are very influential in affecting the chemical composition of plants. *Age* is one of the most important. The percentages of protein, lime and phosphorus are usually highest in the youngest growth. Livestock, including poultry, prefer short, leafy herbage. Therefore, it is essential to keep the vegetation short—not over three inches for bluegrass. Mowing will probably be necessary to keep a range in a condition to supply the right kind of pasture.

CHEMICAL ANALYSES OF SOME NEARLY PURE STANDS, CUT WHEN FOUR INCHES HIGH (R3 – STORRS, 1937)

Percentages in Dry Matter
Species

	Ash	Protein	Fiber	N.F.E.	Fat	Phosphoric Acid	Lime
Ladino clover	8.63	22.95	16.45	48.01	3.96	0.81	2.19
R. I. bent grass	8.44	18.87	20.45	47.99	4.25	0.82	0.99
Kentucky bluegrass	7.27	19.70	21.71	46.86	4.46	0.85	0.95
Orchard grass	8.97	21.21	21.51	42.99	5.32	1.18	1.09
Timothy	7.07	21.58	19.19	47.59	4.57	0.91	1.13

Protein may be 50 percent higher and phosphoric acid and lime 30-40 percent higher in young grass, one-two inches high."

Prof. T. B. Hutcheson, Agronomist, Virginia Agricultural Experiment Station says:

"I would be inclined to recommend perennial pasturage for poultry, and I believe I would make such a pasture with a mixture of 3 to 5 pounds of Ladino clover, 6 to 8 pounds of Timothy, and 3 to 4 pounds of red top or herds grass for this section.

Since Ladino clover does not stand continuous grazing and has to be pastured alternately for best results, I would suggest sowing this mixture and dividing the range into at least three lots, and would run the birds alternately on these lots, allowing the clover to reach a height of 3 to 4 inches before turning on the birds, and then taking them off when they have grazed it down to about 2 inches.

This could be repeated over the entire range; and whenever the forage got too high to be used for poultry, it could be mowed down and allowed to recover and get to the proper succulent stage before the birds are again turned in. Three lots used in this way would furnish pasturage for poultry from about the first of April to the first of November in this state.

As to the use of temporary pastures, I believe one of the best combinations for late fall and early spring use in Virginia is a mixture of 20 pounds of crimson clover and 20 pounds of Italian rye grass to the acre. This mixture may be seeded any time between the middle of August and the middle of September in this section, and should be ready for some grazing in about 40 days after seeding. It, of course, would not give any growth during the winter months, but would come out very early in the spring, giving grazing in most parts of the state by mid March, and would continue to furnish good pasturage until about the first of June.

If 25 pounds of lespedeza is seeded on this crimson clover-rye grass mixture in late February or early March, it will afford pasturage in late summer. Of course, all of these are annual plants, and new seedings will have to be made each year. One bushel of rye to the acre could be seeded with the rye grass and crimson clover, if desired.

In the use of pure cereals, such as rye or winter oats, to furnish early spring pasturage, we would suggest seeding in early September, using at least 2 bushels of grain to the acre. 2½ to 3 bushels would probably be better. Of course, oats may be seeded in late winter, using winter varieties and seeding them in this state between February 15th and April 1st.

Rape may be used either for fall pasturage or for

Crop	Kind or Variety Preferred In Order Named	Amount of Seed to Plant per Acre, Lbs. or Bu. as Indicated	When to Plant	How Soon Available After Planting
Pasture	Lespedeza (common) Carpet grass, Bermuda grass, Clover (white Dutch, Calif. Bur, Hop, etc.)	10-20 lbs.	Mar. to Nov.	3 to 6 months
		4-6 lbs.	Oct. to Nov.	3 months
Chinese cabbage	Any variety	½ lb.	Sept.-March	45 days
Collards	True Ga. White	½ lb.	Any Season	50 days
Kale	Imperial Long Standing Curled Scotch	8 lbs.	August-April	45 days
Mustard	Southern Giant Curled Chinese Broad Leaved	5 lbs.	Sept.-Oct. Feb.-April	40 days
Rape	Dwarf Essex	15 lbs.	Sept.-Jan.	40 days
Carrots	Improved Long Orange, Red Cored Chantency, Danvers Half Long	4 lbs.	Aug. to Mar.	75-90 days
Rye	Fla. or Ga. Black	¾-1½ bu.	Oct.-Nov.-Dec.	50 days
Cowpeas	Brabham Iron	½ to 1 bu.	Mar. to Sept.	50-70 days
Oats	Hastings 100-Bushel Fulghum	2-4 bushels	Oct., Nov., Dec.	45 days
Lespedeza sericea	Sericea	10-20 lbs.	April to July	60-90 days
Millet	Cattail or Pearl	10-20 lbs.		
Peanuts	Fla. Runner	2 bu.	Mar. to June	80-120 days
Napier grass	Elephant, Merker or Napier are all the same	Plant roots or canes	Feb. if roots July if canes	100 days
Swiss chard	Giant Lucullus	5-6 lbs.	Sept. to Mar.	50 days
Lettuce	Chicken Lettuce, Grand Rapids	1-2 lbs.	Oct. to Mar.	50 days
Soybeans	Otootan	¼ to ½ bu.	Mar. to June	60 days

Remarks: Rape may be sown in with oats and rye if rich, moist land is used.
Millet: If for pasture, plant broadcast; if to be cut, plant in rows 3 to 4 feet apart, drilling seed in row.
Napier grass: Set divided root clumps 2 feet apart in 4 to 6 foot rows or stick 3-joint-length canes in prepared ground at 45 degree angle, leaving top joint sticking out of ground.
To get the most succulent and palatable feed, the crops mentioned must be kept green and growing, which means that in most instances liberal amounts of fertilizer must be used, and a succession of plantings of most crops should be made. Most any good truck crop fertilizer applied at the rate of 400 to 600 pounds per acre a week or 10 days ahead of planting, and side or top dressings of quick-acting nitrogenous fertilizer as the plant indicates a need for nitrogen, will be found satisfactory.

very early spring pasturage. When used for fall pasturage, it is usually seeded during the month of August; and when used for spring pasturage, it is seeded after the first of March as weather conditions are found favorable. Swiss Chard should be seeded as soon as frost is out of the ground in the spring, which, in this state, would be about mid March, and repeat seedings may be made up until the first of July. Soybeans may be seeded any time between May 1st and July 15th. The same is true of sudan grass; and for this state, I believe a mixture of soybeans and sudan grass, using about 6 pecks of soybeans and 25 lbs. of sudan grass to the acre, would give better results than either of the crops seeded alone.

I may say that we do not get good results from seeding any of the small grains in the spring in this state, nor do we advise seeding them before September 1st, but either rye, wheat or oats will make good late fall and early spring grazing, if sown between September 1st and October 1st."

The very comprehensive table for the extreme South has been prepared by Messrs. W. E. Stokes, F.S. James and J. Lee Smith in the Bulletin *Green Feed for Poultry in Florida,* directed by D. R. Sowell, Extension Poultryman for Florida.

E. A. Miller, Extension Agronomist, Extension Service, College Station, Texas, says:

"Some good grazing crops to be planted in the early fall, preferably September, if moisture is available, are oats, barley, wheat, rye and rape. Of the clovers we prefer bur clover to Ladino clover because it is better adapted to Texas conditions and especially in the blackland area of Central and South Texas. This crop should also be planted in the early fall and usually furnishes some winter grazing and a good deal of grazing during the spring until it goes to seed. For summer grazing, such crops as Sudan grass, cowpeas, and swiss chard would be suitable.

Unless it gets very dry and hot, these crops would supply grazing during the spring, summer and even into fall. Early plantings could be made from March to June, depending upon the location and later plants could be made during August and September, when the early fall rains set in. In this state, we usually do not have much cold weather until about December except in the northwestern part of the state."

H. Embleton, Poultry Husbandman, University of Arizona, Tucson, says:

"I would suggest earliest planting dates for rye, oats, and barley, September 20; Rape, Swiss Chard, December 1; Soybeans, May 1; Sudan Grass, April 15th. Ladino clover has been tried out here, but was so unsuccessful I would give it no consideration.

In order to get fresh green feed for poultry the year around, alfalfa is sown in the fall along in September. It comes up but does not make much growth until the early spring. From then on it can be depended upon for green feed for $10\frac{1}{2}$ months in the year. During the winter months during which there is very little growth, we have to fill in with barley, wheat, oats, and rye. These are planted in September and tide us over until the early spring, at which time the alfalfa is again ready to cut. One planting of alfalfa will last seven or eight years. During any one year we can get 10 or 12 cuttings."

Prof. B. A. Madson, Head Agronomist of the University of California, furnishes the following regarding pasturing on the Pacific Coast.

"If Ladino clover is available, this will provide green feed for about nine months of the summer,—that is from March through November. Sudan grass in most parts of California can be planted the latter part of April and will provide green feed from about the first of June until frost in the fall, which is the middle of October or early November.

During the winter such crops as oats, barley, etc., would have to fill the gap. Oats or barley, or rye, could be seeded the latter part of September or early October, and will provide green feed within about six weeks or two months after planting; repeated plantings can be made at intervals of two or three weeks to keep green feed coming on as needed. Such crops can be seeded up to early March, so that green feed can easily be provided during the period when Ladino or Sudan are not producing much feed.

Rape and chard can be planted in the early fall— September or early November—and if kept fed down properly, will provide some green feed throughout the winter. However, repeated winter plantings cannot be made with these crops with the same degree of success as you can with the cereals."

The following by E. Y. Smith and G. H. Serviss is from *Cornell Extension Bulletin #502* and tells clearly how to start and maintain pastures in New York.

"Ladino clover, seeded at the rate of at least 4 pounds to the acre, compares favorably with other seedlings for an annual, or temporary, pasture. It produces abundantly when well started, is palatable and nutritious; and may be continued as a permanent pasture if desired and when not overgrazed in the fall.

Rape pastures have been satisfactory as annual pastures. Rape may be seeded early, is palatable, nutritious, and not sensitive to low temperatures. Its disadvantages are that alternate seedings must be made; it grows somewhat too high for chickens, and does not cover the ground well.

Soybeans make excellent pastures during the relatively short growing season in this state. They do not cover the ground well and are somewhat too coarse for best results with poultry.

One ton of ground limestone to the acre should be applied when the soil test shows a Ph of 6 or below.

The seedbed should be well prepared, and the seed sown as early in the spring as the season permits.

Prior to seeding, 600 pounds of 20 per cent superphosphate to the acre should be drilled in or applied broadcast and harrowed in. An application of from 6 to 8 tons of manure (not poultry manure) to the acre is also advisable on poor soils. If manure is not available, an application of 600 pounds of a 0-20-10 fertilizer should be used instead of the superphosphate.

The following seed mixtures have given good results in most sections of the state and on many different types of soil:

1. 12 pounds of Kentucky bluegrass.
   6 pounds of perennial ryegrass.
   2 pounds of Ladino clover.
2. Six pounds of roughstalk meadow grass may take the place of 6 pounds of the Kentucky bluegrass when the seed is available at a reasonable price.
3. Two pounds of wild white clover may be substituted for the 2 pounds of Ladino, but wild white clover is less productive on soils that do not hold moisture and during seasons of low rainfall.
4. Straight seedings of Ladino clover up to 4 pounds to the acre have given excellent results.

The seeding should usually be made without nurse crops.

If nurse crops are used, they should be seeded sparingly and harvested as early as practical. Nurse crops may seriously hinder the new pasture seedings during seasons of low rainfall, but are less harmful when the rainfall is adequate.

To improve old pastures or meadows, ground limestone should be applied as needed; an application of

600 pounds of 20 per cent superphosphate or 600 pounds of 0-20-10 to the acre should be made; and the field should be mowed or clipped frequently during the summer. This treatment usually results in a good wild-white-clover pasture within from two to four years without extra seedings. However, extra seeding with either wild white or Ladino clover, broadcast on the sod very early in the spring, may sometimes be necessary and almost always gives quicker results."

Alvin G. Law, Assistant Professor of farm crops of the State College of Washington sends the following information for the Northwest.

"As you know, the poultry industry is largely concentrated in western Washington so that the green feeds we can grow there are the ones that we are interested in.

We do not use soybeans anywhere in this state as they are not adapted to the type of climate here in the Pacific Northwest. Fall-seeded winter rye, winter oats, and winter barley are used for early spring pasture. Swiss chard and rape spring seeded are used for pasture during the summer and fall and in mild winters through most of the winter."

Every man thinks his own geese swans.
—*Dickens, The Cricket on the Hearth.*

CHAPTER XIII

The End of All Good Geese

Twenty Ways that Geese Are Good to Eat

The goose for the Christmas feast is traditional. Just as turkeys have been the Thanksgiving meat course since the day was inaugurated so has goose held honored place on Christmas for many hundred years all over the world. There is an ever increasing interest in geese and the demand for finely grown and finished young geese is far beyond the supply.

Geese were formerly available only when full grown and during the winter season but now may be had in limited supply young enough to broil during the late summer and fall.

As the demand grows and the breeding and raising of geese is given more attention, as it should, these young geese for broiling should be obtainable at any time of the year.

Geese for roasting are on the market usually from November to January and as the demand grows and improved methods of growing and marketing are adopted by the growers should be on the market at any time.

ROAST GOOSE WITH BAKED APPLES

8 lb. goose	6 or 8 apples
2 qts. bread crumbs	2 teaspoons salt
3 cooked mashed sweet potatoes	Dash pepper
½ cup brown sugar	1 teaspoon sage
	2 tablespoons fat

2 chopped onions

The gizzard, heart and liver should be cooked until tender. Chop and mix with bread crumbs, onion, fat, sage, salt and pepper.

Clean and wash goose thoroughly but do not stuff.

Prick through skin into fat layer around legs and wings. Heat in moderate oven (375° F.) for fifteen minutes. Cool to room temperature and repeat twice more. Drain off fat, rub inside of goose with salt, stuff and sew. Place in roaster and cook uncovered in slow oven (325° F.) until tender—about 25 minutes per pound. Wash and core apples, sprinkle with brown sugar, stuff with seasoned sweet potatoes and place in pan with goose one hour before goose is done. Serve hot. Serves eight people.

GOOSE ORANGE SALAD

2 cups dice cooked goose	½ cup french dressing
2 cups diced celery	Lettuce or chicory
2 cups sliced oranges	¾ cup toasted almonds

Combine first four ingredients. Mix and let stand for one hour. Drain and let stand for one hour. Drain and arrange in salad bowl lined with lettuce or chicory. Garnish with almonds. Serves six.

ROAST GOOSE STUFFED WITH SAUERKRAUT

Omit bread stuffing. Stuff two quarts drained sauerkraut into goose. Roast in slow oven about 325° F. until tender, about 25 minutes to the pound. Apples and sweet potatoes may be omitted.

STUFFED GOOSE NECKS

Goose meat (uncooked scraps)	Goose neck skins removed whole
Bread stuffing	1 onion sliced
1 cup hot water	

Grind finely scraps of goose meat and mix with stuffing. Tie one end of skin with string and fill with mixture. Tie other end and place in oven, add onion and water and bake in moderate oven (350° F.) until brown and crisp, basting occasionally. Serve hot sliced.

PATE DE FOI GRAS

½ cup cooked goose livers	Salt and pepper to taste
2 tablespoons goose fat	⅛ teaspoon paprika
3 hard boiled eggs	½ teaspoon grated onion.

Saute livers in goose fat until tender. Mash into a paste with eggs. Add salt, pepper, paprika and onion. If too stiff add additional goose fat. Spread on thin slices of toast. Makes ½ cup of paste.

GOOSE LIVERS IN JELLY

4 prs. goose feet	⅛ teaspoon pepper
4 goose livers	1 teaspoon salt
1 cup water	½ clove garlic
	1 small onion minced

Clean, scald and strip the skin from feet. Cover with water, add salt and cook until bones fall apart, about two hours. Add more water as it evaporates, keeping original amount. Strain and boil liquor down to one half its volume. Combine onion, garlic, salt, pepper and water and cook for fifteen minutes. Add livers and let simmer twenty minutes adding more water to keep livers covered. Grind livers with onion and garlic, using a little of the liver liquor to rinse the grinder and add it to the ground livers. Half fill small molds with liver mixture and fill with stock made from boiling feet. Chill. Use as appetizer. Serves 12. A thin slice of lemon may be floated on top of mold. A small loaf pan may be half filled with liver and filled with stock and a design of lemon slices and strips of red and green peppers floated on top. The loaf may be cut into squares. Use hollowed tomatoes for molds. Chill one cup of the stock until firm. Dice or break with a fork and use as garnish for tomatoes.

GOOSE GRIEBEN (Cracklings)

Rendered fat skin of a goose	Salt
1 cup cold water	

Cut the fat skin of goose into small squares, sprinkle with salt and let stand 12 hours. Wash and drain. Add cold water, cover and simmer for about an hour. Add water to prevent fat from browning too fast. Drain and fry slowly. Careful not to scorch. Brown well and place in oven for a few minutes. Drain on brown paper.

ROAST GOOSE AND CABBAGE

8 to 10 pound goose	3 heads finely chopped cabbage
3 sliced onions	
Salt and pepper	

After preparing the goose for roasting, divide into serving portions. Place in pan and roast in slow oven (325° F.) until almost tender. This will take from two to three hours. Drain off most of the fat into a pan; Saute onions in the fat until golden brown and then add cabbage. Cook 5 minutes then season.

Lay the portions of goose on top of cabbage and sprinkle with salt and pepper. Then cover and let the whole simmer for an hour until the goose is tender and cabbage well cooked. Serves eight people.

ROAST GOOSE AND BAVARIAN CABBAGE

Follow instructions as above but leave goose whole. Omit onions and cabbage. Drain off bulk of fat and then cover bottom of pan with Bavarian cabbage. On this place the goose and roast until tender.

BAVARIAN CABBAGE

6 cups shredded red cabbage	1 cup boiling water
2 tablespoons goose drippings	3 tablespoons flour
3 whole cloves	⅛ teaspoon cinnamon
2 sour apples	4 tablespoons brown sugar
Salt and pepper	2 tablespoons vinegar

Heat goose fat in skillet and add cabbage, salt, pepper, cloves and peeled and quartered apples. Add boiling water and cook slowly for about 20 minutes. Mix flour, cinnamon, brown sugar and vinegar and add to cabbage.

FRICASEED GOOSE

Back, wings, neck, gizzard and heart of the goose	Clove of garlic (minced if desired)
1 teaspoon chopped parsley	½ sliced onion
Boiling water	2 tablespoons flour
2 celery stalks	Salt and pepper
1 cup goose broth	Dash of ginger

Rub meat with salt, pepper, ginger and garlic and let stand 12 hours. Then add celery and onion, cover with boiling water and let simmer until tender. Mix flour with goose broth adding a little at a time to make a smooth paste. Stir while cooking, until thickened. Boil this about five minutes. Add parsley and pour over goose. Serve with dumplings. Will serve eight.

FRIED GOOSE LIVERS

Remove gall and place liver in cold salt water for an hour or more. Wipe dry with clean cloth. Brown in hot goose fat.

GOOSE LIVER CROQUETTES

Enough milk to soak up two
slices of dry white bread
2 slices bread
4 eggs beaten slightly
½ teaspoon salt

1 tablespoon minced parsley
Pinch nutmeg
6 tablespoons goose fat
1 cup chopped goose livers
4 cups goose stock

2 tablespoons flour

Crumble bread and soak for four or five minutes in just enough milk to saturate. Add eggs, parsley, salt and nutmeg. Melt 4 tablespoons goose fat, add bread mixture and cook for four or five minutes over very low heat until it thickens. Stir constantly while cooking. Add liver and cool. Shape into croquettes and cook covered in hot stock about one half hour. Remove from stock and blend remaining goose fat and flour. Add stock and cook until thickened and pour over croquettes. This will serve four.

GOOSE JAPANESE

8 pound goose
4 cups water
⅔ cup brown sugar
2 tablespoons salt

½ teaspoon cinnamon
½ teaspoon white pepper
½ teaspoon sage
½ teaspoon allspice

½ teaspoon cloves

After cleaning goose combine remaining ingredients in kettle large enough to hold the goose. Heat to boiling point and cook about 12 minutes. Place goose in the liquid, cover tightly and let simmer until tender, allowing 20 to 25 minutes per pound of goose, basting occasionally. Remove from heat and let stand twenty minutes before serving. Serve with gravy for ten.

SALMI OF GOOSE

4 tablespoons fat
1 cup sliced onions
2 cups canned tomatoes or
tomato juice
1 teaspoon salt
1 green pepper diced

2 whole cloves
1 bay leaf
½ tablespoon brown sugar
3 tablespoons flour
¼ cup cold water
Sliced cooked goose

Saute onions in fat until tender and add remaining ingredients except flour, water and sliced goose. Cook slowly for fifteen or twenty minutes. Strain and add flour mixed with water and cook till thick. Then add sliced goose and heat through. This will serve six.

BRAISED GOOSE BREAST AND LEGS

The breast may be used with or without skin. Brown in a small amount of hot fat. Add a little boiling water, then season, cover tightly and let simmer until tender. Serve hot with applesauce.

DEVILED GOOSE

8 pound goose	2 tablespoons prepared
Stuffing	mustard
Boiling water	1 teaspoon pepper
1 tablespoon salt	¼ cup vinegar

Clean goose, cover with boiling water and let simmer for 1 hour. Drain well and wipe dry. Fill the neck and body partially with stuffing, sew up and truss. Roast in slow oven (325° F.) allowing about 20 minutes per pound. Mix pepper, vinegar, mustard and salt and use to baste goose. Serve with giblet gravy. This will serve six to eight people.

SMOKED GOOSE   ●

Meat from wings, neck and	Neck skin left whole
back of goose	1 tablespoon sugar
Breast and legs	1 teaspoon salt-peter
⅔ cup salt	½ clove garlic

Scrape meat from neck wings, neck of goose and chop fine. Stuff into neck skin and tie both ends. Rub filled neck, breast and legs with salt. Mix remaining ingredients, rub over goose meat and place in crock with clean cloth. Weight down with plate and set in cool place for seven days turning occasionally. Drain, cover each piece with cheese cloth and smoke. Chill, slice thin and serve.

HOW TO SKIN A GOOSE

Always very popular with housewives, wise in culinary arts and science, goose fat has been a standby for many different purposes since records of cooked dishes have been kept and food recipes swapped and handed down from one generation to the next. It is in a class by itself for peculiarly delicious flavor, low melting point and accessibility. Skinning, or as the process was called in olden times in Europe "shinding" is the best process of preparing goose fat for use.

The most delicately flavored fat is that laying next to the skin and so geese are bred, raised and fed to increase not only the amount of fat but to improve the flavor and texture of the flesh beneath the layer of fat just under the skin.

Take the fat goose when plucked and singe and wash thoroughly with a solution of baking soda, using a soft brush, cloth or sponge.

Wrap in a clean linen cloth or put in a paper bag and chill or freeze so that the fat will be hard and firm.

TO SKIN

Cut the skin down the backbone from neck to tail-head. Cut around the first joint of each leg and around the juncture of wings with body and clear around the base of the neck. Then separate the layer of fat from the flesh underneath with a small dull knife. Lift the layer of fat and skin with the left hand while with the right continue to separate the fat from the flesh.

The connective tissue is very thin and tender in some places while in other places care must be taken not to cut into the meat as the fat is practically attached to the flesh. Cut the pieces of fat off as they become too large to hold handily and drop them into a kettle of cold water with from a pint to a quart of water depending on the size of the goose and amount of fat. There should be just enough water to completely cover the fat when finished.

Then heat the water to boiling point and reduce heat and let simmer until the water has completely boiled away. Cool and pour fat into clean jars or cans. Store in a cool place.

The remaining cracklings may be further reduced until brown and this fat should be stored separately and reserved for meat cookery as it has a "browned" flavor.

After the skin has been removed, break the body at the joint below the ribs and remove the entrails. Carefully remove the envelope and the fat that surrounds the giblets. This is rendered separately and used in meat cooking when clear fat is required. In all cases place a slice of raw potato or fresh bread in the rendering kettle which tends to carry off any odor or flavor.

The cracklings, well salted, are called "grieben" and used as appetizers.

HOW THE FRENCH PRESERVE GEESE
By Pierre J. Menville, Chef at Macy's

BASSES PYRENEES STYLE

After the geese have been fattened they are well plucked and cleaned of their small pin feathers. Then cut into four pieces. Take the whole quarter back in its full size. Split the bird in half to cut the wings. Then take off the breast bone and spine. When this is finished weight the four quarters together. Then salt—using 1 pound of salt to 5 pounds of goose. Rub salt in well and put 3 to 4 grains of whole black pepper and 1 clove of garlic on the more fleshy part. To do this—take a pointed knife and insert into the flesh. Raise the handle and push the peppers and garlic through the hole. Remove knife and press down. Now take this prepared goose and place on a board which has been covered with clean straw. Let stand for three days. Then salt again, but use less salt. Let stand for three more days. Now the meat is ready for preserving. Take the goose fat that has been removed and cut into small pieces. Place in a deep pot or dutch oven. The dutch oven is preferable. Do not use an aluminum pot. Put geese in pot, being certain to brush off all surplus salt. Melt fat slowly and fry meat at same time. When fat is well melted and meat is nice and brown and well fried (not too dry) take off the fire. Drain the grease into a crock jar. The best way to do this is to strain through a clean piece of burlap. Then wait 1 hour or more until the grease just begins to set. Take the meat and place nicely, so that it will be covered by the grease. If there is insufficient grease to cover use some other grease or even Crisco. When the fat has congealed, cover with wax paper and keep in a cool, dry place.

To use—take a piece of the meat from crock and warm up and you then have your roasted goose.

The goose grease itself is a fine delicacy when spread on bread in place of butter.

> The lazy geese, like a snow cloud
> Dripping their snow on the green grass,
> Tricking and stopping, sleepy and proud,
> Who cried in goose, Alas.
> —*Bells for John Whiteside's Daughter*
> *by John Crowe Ransom.*

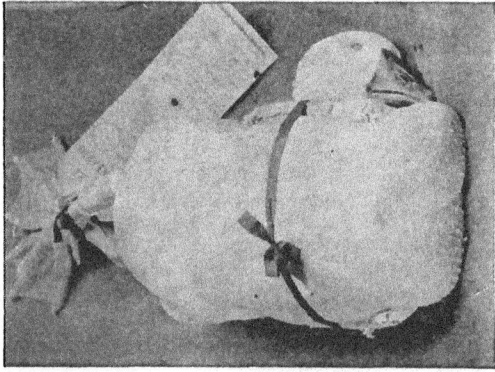

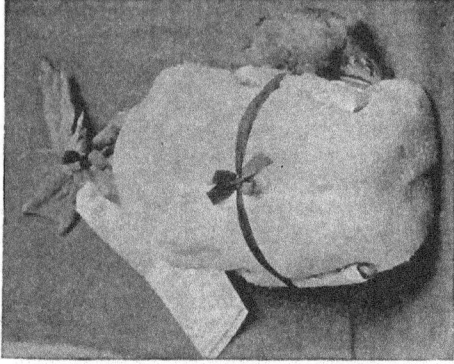

"Well fattened and well dressed." Courtesy Franklane L. Sewell.

PART TWO

DOMESTIC DUCKS

"The manners and actions of the duck, whether upon land or water, are curious and pleasant to contemplate. Their regular afternoon parade and march in line; the elder drakes and ducks in front, from the pond homewards, is a beautiful country spectacle, to be enjoyed by those who have a relish for the charms of simple nature."

—*Bonnington Moubray, Esq., A Practical Treatise on Domestic Poultry, 1819.*

PREFACE

As stated in the introduction, this book was originally planned to tell the whole story of the goose which has from the very first interest in poultry been so generally neglected. But as I proceeded with the subject and studied the situation from many different angles, I found that many breeders of geese and other folks interested in them, who were so situated that they could be profitably kept, were also in a position to pleasantly and profitably include ducks in their waterfowl enterprises and so would welcome authentic information on ducks within the same covers that contained the story of the geese.

Especially did this seem feasible and rather important when I found that nothing had been written and published in book form on ducks since 1920 and that the whole situation regarding the raising, breeding and marketing of ducks has so materially changed in twenty-five years.

New breeds have been developed and new methods of care and management have come into use. The marketing situation has changed to a marked degree, artificial incubation has been perfected, studies of feeding and disease control have added greatly to satisfactory duck culture; and perhaps the strongest factor in my decision to include ducks in this book was a talk I had with Mr. G. E. Eiermann, President of the Orange Judd Publishing Company, New York, who told me of the marked and increasing demand for more reliable information on ducks; since the last book published on ducks and geese is now out of print as well as out of date.

As in the first section of this book, devoted to geese, I shall endeavor to make this an authentic and comprehensive coverage of the subject in layman's language and with as few technical terms as possible.

The information in this book is from writers of whose knowledge and credibility there can be little doubt and from the personal experience of the writer covering many years.

It has been impossible to trace the growing of ducks from ancient times as in the case of the goose, for the reason that the goose was in olden times far more important than the duck—and even than the domestic fowl or chicken, if we go back far enough into antiquity.

Methods of goose raising have differed very little from ancient days and duck raising has advanced to an entirely different enterprise which, with little or no data on early management of ducks, makes it almost impossible to trace the changes.

However, with the trend to country living and part-time and subsistence farming on small suburban home lots and little farms, there is a growing demand for information on all kinds of livestock breeding, growing and management and of all the numerous kinds of birds and animals, none fit into such plans better and with less trouble and expense than our web footed friends both ducks and geese, and this book I trust may be of service to these folks for whom there is really no complete source of information at the present time.

As an interesting hobby or a profitable source of supplementing the family income by producing eggs and meat for the table, the duck has a very real and useful place. For the country estate fortunate in having stream or pool or lake, ducks on the water complete the picture and make an always interesting sight at a minimum of trouble or expense. A flock of any breed of ducks in its native element is most attractive. The grays, browns, blues, greens and clarets of the Mallard, the iridescent green-black of the East Indian, the beautiful tiny Calls or a flock of White Crested, snow white in color and with their large, globular crests like snow balls, all have their appeal and any one, or more, of these breeds, or many others, may be selected to suit the owner's taste.

There is ample evidence by hundreds of duck farms scattered over the Atlantic coast and in the mid-west that the growing of ducks can be made a steady and profitable business and for the fancier who is interested in the breeding of fine specimens for the sport of competition in shows and expositions, there is a growing interest in fine ducks of all the various breeds and varieties.

So, to all these different classes of folks who are looking for information, I shall try to make this book valuable as an authentic source of the knowledge they need.

The self-sustaining Muscovy. This White Muscovy Duck laid 20 eggs, hatched and raised 20 ducklings on the Government Experimental Farm at Ottawa, Canada.

CHAPTER I

Ducks

Ducks belong to that general order of birds designated as Antidae, a class of waterfowl which also includes geese and swans. While this family includes geese, as well as ducks, yet ducks are more aquatic in their habits than geese and are omnivorous in their diet, eating grain, green food and animal matter freely. As a class, they are considerably smaller than geese, possess broader and flatter bills, shorter necks, and a lower station which in a measure accounts for their awkwardness upon land. They are further distinguished from the goose tribe by the fact that the males of the colored varieties are usually brighter colored than the females and show little, if any, inclination to assist in the brooding and protection of the offspring; while ganders take a very useful part in the care of the young goslings.

Lewis Wright in his *Illustrated Book of Poultry* (London 1872) states "We believe all Naturalists are agreed that the whole of what may be called the 'farm' breeds of ducks, if not many of the others also, are descended from the Wild duck or Mallard (Anas Boschas) which is distributed more widely perhaps than any other bird over the entire continent of Europe and a great part of North America. Indeed it's range may be stated to extend from the vicinity of the pole in summer to almost the torrid zone in winter, migrating regularly towards the south on the approach of cold weather and returning with summer to the more northern regions.

"In the more southerly countries it is, however, a less frequent visitor, the temperate latitude being its favorite home; it has been known to reach even North Africa during its winter migrations; Italy, Greece and Spain are its most favorite winter quarters.

"The color of the wild duck (Mallard) nearly resembles that of the Rouen to which we may therefore refer for more detailed description as far as this point is concerned.

"The shape however, is more slender and upright and the habits much more active as shown in the very excellent presentation on another page which also most clearly shows the resemblance in other points. . . . But a singular change in the plumage of the drake must be here noted which is common also to the Rouen (as shown on page 185) and to most other varieties of ducks in which the plumage of the male is greatly superior in beauty to that of the female. It is thus described by Waterton:—

" 'About the 21st of May the breast and back of the drake exhibit the first appearance of a change of color. In a few days after this the curled feathers above the tail drop out and grey feathers begin to appear amongst the lovely green plumage which surrounds the eye. Every succeeding day now begins marks of rapid change.

'By the 23rd of June, scarcely one green feather is to be seen on the head and neck of the bird. By the 6th of July every feather of the former brilliant plumage has disappeared and the male has received a garb like that of the female though of a somewhat darker tint. In the early part of August this new plumage begins to drop off gradually and by the 16th of October the drake will appear again in all his rich magnificent dress; than which scarcely anything throughout the

Upper—Rouen Drake showing summer plumage. At this season the
Rouen drake assumes a plumage resembling quite closely that of the
female. In the fall the drake again assumes the normal male plumage.
Lower—Rouen Duck. (Photographs from the Bureau of Animal Industry, U. S. Department of Agriculture.)

whole wide field of nature can be seen more lovely or better arranged to charm the eye of man.'

"The dates here given are subject to some little variation as Waterton observes, but much less so than one would suppose . . ."

"Wild ducks have often been domesticated. The usual mode is to obtain eggs and hatch them under tame ducks or hens, when they are brought up with no difficulty, though of course there is some wildness of disposition. It is also generally found that down to even the third or fourth generation such domesticated wild drakes pair strictly like their ancestors instead of taking a small harem like the domestic variety. When thus domesticated the progeny after a while almost always begin to vary in color . . ."

In this color variation we find the embryonic beginning of the mutations which finally result in all the varieties of domestic ducks which most writers and naturalists, as Wright says, agree, probably stem from the Mallard, with the exception of the Muscovy which is a distinct species as proven by his inability to produce fertile offspring when mated with any other variety. All other breeds and varieties of ducks interbreed freely and the offspring breed as freely.

When the goose drinks as deep as the gander, pots are soon empty, and the cupboard is bare.
—*C. H. Spurgeon, Ploughman's Pictures.*

CHAPTER II

Breed Standards for Ducks

Many years ago when the author first became especially interested in the production of ducks and their breeding, reliable information concerning typical shape and color requirements for the several breeds and varieties was not only most difficult to discover, but very slight when found. Indeed, even the Standards of that time were comparatively crude and indefinite. So little attention in fact, was paid to waterfowl at shows that poultry judges on occasions, refused to pass upon the duck classes, leaving that task for the show secretary or some other official of the organization.

Since then, conditions have improved but little and literature pertaining to ducks is still scarce and rather casual. The revised *Standard of Perfection,* although still far from perfect, discloses some improvement over previous editions and indicates a fuller appreciation upon the part of the American Poultry Association as well as practical breeders, of the importance of uniform and Standard quality in waterfowl. It should be stated, however, that the fact waterfowl breeders have any Standard at all is due to the broad-mindedness of the chicken fanciers rather than to their own efforts. The American Poultry Association was brought into being by chicken breeders as distinguished from waterfowl fanciers and these "chicken men" have always assumed the initiative in the matter of breeder's cooperation so it is much to their credit that waterfowl have been given the liberal consideration which has been be-

187

stowed upon them in the making of their breed Standards.

Waterfowl breeders as a whole have proven most apathetic in their efforts to standardize their breeds, and as a result have failed to lend the Standard Revision Committees of the American Poultry Association at the periodic revision of the Standard the necessary assistance in the writing of suitable descriptions of the various breeds; therefore, any shortcomings which may appear in the waterfowl section of the various editions of the Standard can not fairly be charged to the Standard Revision Committee, but to the indifference of the waterfowl breeders themselves. In all fairness, the Standard Revision Committee is to be commended for what it has accomplished in the face of the indifference encountered while revising the Standards.

However, with the newly reorganized and vigorous American Waterfowl Association and its broad, aggressive and intelligent program for the betterment of the whole American Waterfowl industry and fancy, we may expect adequate cooperation with the Committee on Standards of the American Poultry Association resulting in more complete descriptions, more illustrations of the breeds already admitted and the admission of several worthy breeds not now included in the *Standard of Perfection*.

Of the literature which has in the meantime appeared on waterfowl subjects, it may be said that what little there has been has dealt more particularly with the commercial phase of the industry and such attention as has been bestowed upon the matter of breed requirements has been for the most part rather general in its nature. The result is many waterfowl breeders are still handicapped with a lack of detailed information pertaining to their pursuit, while breeders and judges alike complain that they find it impossible to secure definite data covering the shape and color requirements of even the more popular of the few breeds of ducks and geese now recognized.

It was partly a realization of the seriousness of this situation that prompted me to begin this work with a view of supplementing the little information now available with such additional details as I have been able to collect from dependable sources in both this country and Europe. If my efforts should bring about a clearer understanding of the requisites of the various breeds of domestic ducks, I shall feel that my labors are well rewarded.

In passing, I wish to acknowledge both my indebtedness and sincere gratitude for the valuable assistance so freely rendered me by the duck breeders of this and other countries whose hearty cooperation enabled me to complete this volume.

Much of the confusion and differences of opinion which have attended the interpretations of the descriptions of the different breeds of ducks by breeders and judges alike, are due to the loose, careless and very superficial phraseology employed in the writing of Standards. Often, the terms used are actually contradictory and therefore null and void. For example, the description "long, broad and deep" is frequently encountered. All three adjectives are merely relative terms and no one of them conveys any meaning, except in comparison with other dimensions such as width or depth. But when in describing surfaces the term broad is employed in conjunction with long, no mental picture can be formed since a contradiction is set up. If a thing is thought of as long, it is at the expense of width and on the other hand if it is conceived of as broad, it is in contrast to its length. It is clear, therefore, that a surface especially when compared with itself, cannot be both long and broad, and such descriptions as a result, instead of being clarifying are, if anything at all, confusing; and when anything is described as "long, broad and deep" what is one to think? Four breeds of ducks in the present *Standard of Perfection* are described as "broad and long" and three as "broad, long and deep." There is urgent need for more specific and definite terms in describing the body shape as well as descriptions of other sections of the duck, both in shape and color, in the present Standard.

For, as a matter of fact it was the widespread lack of agreement upon these very points which prompted the adoption and the publication of the breed Standards as they are found today, inadequate and indefinite as they may be. Indeed, it is for the purpose of educating the poultrymen of this country upon the subject of breed character that Standards are published and offered to the breeders of America and that such standards fall short of their objects must be very plain to all who will give but a moment's reflection to this matter.

Standard weights have been adopted for the reason they have been proven best adapted for the preservation of the peculiar characteristics of the respective breeds to which they apply, developed as they have been each to fulfil its particular mission in life and are given later on with the descriptions of the respective breeds.

In spite of the very general temptation to breed for greater size, it will be found that each breed will prove more satisfactory in its particular sphere if the Standard weights are adhered to. Generally speaking, both fertility and fecundity diminish as the size increases, and too frequently vitality is also sacrificed along with other desirable qualities by the worshippers of avoirdupois. The Standard weight for each breed is usually the weight that will be most useful. Especially in selecting specimens for exhibition purposes the fact should not be overlooked that when other points are equal, judges are required to place the Standard weight specimens over those exceeding that figure. Judges who over-emphasize weight in waterfowl are subject to censure by organizations and individuals having the best interests of the waterfowl industry at heart.

On the other hand, undersize in waterfowl is even more objectionable except in certain small breeds where very small size is desirable. Very small individuals in other varieties usually show lack of proper care, under-nourishment (both of which result in low vitality) or poor breeding—or all three combined. When the result of any or all of these causes, such specimens are not desirable for breeding purposes and should be culled out of the flock. All meat breeds of waterfowl should be *reasonably* fat when exhibited, to appear at their best, and size which will permit Standard weights under such conditions. may be considered typical for the breed. Size which requires excessive fattening to attain Standard weights may be considered inadequate. In the ornamental and egg breeds a good condition of flesh only is desirable for the show room, therefore, a size which will make Standard weights without fattening is correct.

Still the fancier and commercial breeder, alike, should constantly bear in mind that size is but one of the requisites of any breed of ducks. Typical shape, correct color and indications of usefulness are more important than size and the real breeder is the one who devotes as much attention to one of these as to the other as they are all interlocking virtues and either one by itself is of little value. What is desired and should be the aim of every breeder is the specimen well balanced in all essential qualities.

In the breed descriptions which follow, I shall endeavor to avoid the shortcomings of the conventional and meaningless descriptions previously discussed. Wherever it is practical in outlining the shape sections, the ratio in which the length of the section exceeds the breadth will be generally stated. It must be borne in mind, nevertheless, that these proportions have not received official endorsement of any club or association, but are merely the results of observation and in some cases of careful measurements of certain outstanding specimens observed at the leading shows and in the more successful breeder's yards. Great care has been taken in specifying typical proportions in each breed and I feel sure that these will not be found far from correct.

> "On Candlemas day
> Good Housewife's geese lay
> On St. Valentine
> Your geese lay, and mine"
> —*Rev. Edmund Saul Dixon in Ornamental and Domestic Poultry, 1848.*

CHAPTER III

General Nature—Various Types—Adaptability of the Different Breeds of Ducks

In a state of nature, ducks are monogamous and with but few exceptions deposit their eggs in nests built upon the ground, concealing the eggs carefully, by a covering of grass or leaves, each time the female leaves the nest. Under domestication, the drakes soon turn partially and often wholly polygamous and after a comparatively few generations of civilization the quality of conjugal fidelity practically disappears. In most of the domestic breeds the nest building instinct has also been forgotten and the eggs are more likely than not to be laid carelessly about the floors of pens or on the ground; although just before becoming broody (in case broodiness develops) a nest is sometimes built and the eggs laid there in nature's way.

Ducks thrive well in restricted quarters and lend themselves more readily to intensive culture than do other domestic waterfowl. As a consequence duck breeding has for a considerable period been an important industry in such thickly peopled localities as Belgium, northern France, parts of Ireland and England, China and Japan.

Up to about 50 years ago, ducks in this country were not generally considered a profitable fowl to raise. There was little demand for them but modern feeding and raising methods have so improved the quality of the flesh that a heavy demand for them has gradually developed. To satisfy this demand large duck raising establishments have sprung up throughout the Atlan-

General view of water yards and ducklings on a large Long Island duck farm. (Photograph from the Bureau of Animal Industry, U. S. Department of Agriculture.)

tic and mid-west states, many with annual capacities of from twenty to one hundred thousand ducklings which are marketed as "green ducks" at two and three months of age.

Notwithstanding the production of "green ducks" on these vast commercial duck breeding establishments, the majority of ducks in this country are produced in small flocks upon the farms of the middle west. However improbable it may sound, it is nevertheless a matter of record from government reports that both the states of Iowa and Illinois lead any of the eastern states in this respect.

Duck raising is not confined solely to the production of ducks for food. In this country, in a very small way and in Europe to a much greater extent, ducks are also kept for egg production. Some breeds have been developed to a point where they lay quite as well, and in some instances better, than hens and there is no basic reason why they should not prove as profitable egg producers as hens. Other breeds are bred for purely ornamental purposes and some are produced for decoys to be used by hunters for attracting wild ducks within gun range.

A flock of ducks can be profitably raised each year upon most farms and where range is available will glean most of their livelihood from the insects, waste matter and grass about the farm. In view of the fact that a small percentage of farms are engaged in the rearing of ducks, it seems that here is plenty of room for the development of a profitable side line on innumerable farms in this country.

The various breeds of ducks may be classified according to the spheres of usefulness to which each is best adapted, into the heavy or meat breeds, the general purpose ducks, the laying class and finally, the ornamental group. In the first class is included such breeds as the Aylesbury, Pekin and Rouen; in the second will be found the Blue Swedish, American Buff, Cayuga, Crested White and the Muscovys; the laying class is represented by the three varieties of Runners and the Khaki Campbell and the ornamental division comprises the Black East Indias, the Calls and the Mallards as well as a number of non-standard semi-wild ducks, which are not included in this book.

Of the meat breeds, the Pekin is almost universally raised in the United States commercially, owing to its being best adapted to the intense production of "green ducks" upon the great business duck plants along the Atlantic coast and in the middle west. In England, the Aylesbury is esteemed quite as highly for identical reasons. Although both are first class breeds for the general farm flock, still they do not, under farm conditions, reveal the marked superiority universally conceded them over the darker colored breeds upon the large commercial plants.

Farm raised ducks are usually not sold until past the pin-feather age, so the dark pin-feathers which are so objectionable on young "green ducks" do not prove such handicaps. Then, too, many of the farms are without water for bathing purposes and it is under conditions such as this that the white breeds are at a distinct disadvantage on account of the proneness of the white plumage to show stains.

The best examples of the general purpose class, if such a classification is proper among ducks, are the Blue Swedish, American Buff, Cayuga and the Muscovy. These ducks most nearly correspond to the so-calle1 general purpose class of chickens, being not only medium sized and of excellent table qualities but much better layers than the heavier ducks. Indeed, in egg production these breeds will probably excel the average chicken in the general purpose class and while not quite as profitable "meat ducks" as the Aylesbury, Pekin or the Rouen on account of their smaller size, still this shortcoming is more than compensated for by their better breeding and hatching qualities.

The three Runner varieties and the Khaki Campbells are the representatives of the extreme egg type. They are in the class with the Chinese among geese and Leghorns among the breeds of chickens.

While America has never favored the duck as an egg machine to the extent England has, there is no doubt that under certain conditions the keeping of laying ducks should prove fully as profitable, if not more so, than the keeping of laying hens. Oscar Grow writes "Ducks can be successfully kept where it would be futile to attempt to raise chickens and, as they require less elaborate quarters and are virtually free from both vermin and disease, it would appear that duck eggs can be produced more economically than hen eggs.

"For home consumption duck eggs are as desirable as those from hens, and some authorities go so far as to claim that for culinary purposes, duck eggs are much superior to the more common chicken product. If this be true, it may be assumed that as the merits of duck eggs become better known, a keener demand for them will develop and as a consequence the egg duck should become correspondingly more popular."

The ornamental breeds, aside from the Crested duck, have little to commend them for purposes other than as pets and ornaments, although the Grey Call is used considerably by hunters for decoys. The Crested Duck possesses practical value in addition to its quaint beauty. It is large enough to make a fairly good table duck and breeds and hatches well. However, inasmuch as it is bred chiefly for its novel appearance it properly, therefore, belongs in the ornamental class.

The East India duck cannot claim recognition except for its ornamental virtue but its black plumage high-lighted by the lovely shimmering green sheen makes it universally admired. The Mandarin and the Wood Duck, both of which are not standard breeds, are purely ornamental in character.

A traveller at Sparta, standing long upon one leg, said to a Lacedemonian, "I do not believe you can do as much." "True," said he, "but every goose can."

—Plutarch, Laconic Apothegms.

CHAPTER IV

Breeds of Ducks—Origin and History

Meat Breeds

The Aylesbury Duck

"This is the best of all table breeds; very large, white fleshed, getting its name from the vale of Aylesbury in Buckinghamshire where the breed originated and where it is still reared in vast numbers each year for the London market." Thus writes Reginald Appleyard, English waterfowl authority in his book *Ducks* published in England a few years ago: and this seems to be the opinion of most English writers on poultry and the duck raisers of England as well.

Of its origin, Mr. Harrison Weir in *Our Poultry* published in 1902 states: "Having heard that a Mr. Charles Ambrose of Ely was said to have a flock of White ducks descended from white Wild Mallards, I wrote him for the facts and received the following reply: 'On our fen farm we bred a great many duck and Mallards, and about seven years ago I managed to get a white drake (Mallard) in the same way that a white sparrow or black-bird or starling occurs now and then. I put this, pinioned, with two Mallard ducks and bred from them, getting only two percent white the first year. This year (1899) I have got them to breed all pure for the first time; but during the previous years they did not breed any odd colors: they were either Mallard color or pure white. I have now two hundred pure white.' "

Robinson in *Growing Ducks and Geese for Profit*

Upper—Young Pekins which on account of their size, thriftiness and rapid growth were selected out of a lot about to be killed for market and saved for breeders. Lower—Aylesbury Drake—Notice the depth and development of the breast. (Photographs from the Bureau of Animal Industry, U. S. Department of Agriculture.)

commenting on this incident says: "Here we have an instance of making a domestic white variety of the Mallard in a short term of years in recent times. Both Weir and Brown (Sir Edward Brown, author of *Races of Domestic Poultry*) rightly cite this as showing that the Aylesbury ducks were—in all probability—produced in a similar manner at some unknown time in the past. Such an instance as this, with only wild blood used, is of incomparably more value for the light it throws on the probable evolution of white varieties of poultry in ancient times than the cases of the White Plymouth Rocks and White Wyandottes. For in those cases, both appearing in recently made breeds of very much mixed origin, there is always the question whether the white specimens appearing are from ancestry containing no white individuals or are cases of reversion to an ancestor not far back in the family lines."

With all the evidence pointing to the Mallard as the distant ancestor of the Rouen, it seems very likely that the Aylesbury, which aside from color is almost identical with the Rouen, descended in like manner from a white sport of the same family.

While the Aylesbury has never been popular in the United States for market purposes owing to the general belief here that it is a slower grower than the Pekin and not as vigorous, it still continues to lead all breeds as a market duck in the land of its origin.

But, even there with the Aylesbury ducks, as it is in our own country with a number of breeds of poultry, there are two types bred; one for exhibition and one for "utility." There, fad breeding for show purposes, has produced the exhibition type with greatly overdone keel and tremendous body, slow growing, sluggish in temperament and low in vitality, fecundity and fertility. Appleyard states further that with the big "keely" type of ducks "a useful mating is one drake with two or three ducks but with the lighter, more active utility type five or six ducks may be run with one drake."

"The true Aylesbury," he continues, "is just a moderate layer and if of a good strain will generally give a

A pen of English Aylesburys from the yards of Reo Mowrer, Unionville, Mo.

flock average of a hundred eggs or so in a year. Remember, they are bred for meat, not high egg records."

Aylesburys were shown in this country at the first Boston Poultry Show held in 1849 by John Giles of Providence and at the Worcester Fair the following year a Mr. Lincoln exhibited an Aylesbury among other breeds of ducks. From then on, they have been bred and shown in this country but have never been popular as a commercial duck for without high, economic values, no breed of ducks except the purely ornamental varieties ever has, nor can ever hope to be, and stay, popular or be very widely bred. So, while in

England the lightweight type of Aylesburys have pretty well headed the field as business ducks, the Aylesbury has never been any serious contender with the more active and virile Pekin for popularity in the United States.

The Aylesbury as bred to our American Standard is a striking and impressive breed. It has the appearance and the temperament of other highly specialized meat animals. The same general type and apparent temperament as the Toulouse goose, Dorking fowl and even the highly developed Hereford beef cattle.

The large head, long straight bill, full bold eye, long slender neck, very slightly curved; big, deep body carried nearly horizontally, straight keel running back from breast to stern, straight, flat back with strong wings folded closely to the sides gives the effect of strength and solidity and that here is *meat* in a compact and bountiful package grown on a frame built for the purpose.

The eye should be dark and full, the bill, flesh color or even pinkish; the legs and feet bright orange and the plumage, unlike the creamy white of the Pekin, should be a pure pearly white.

There will always be ardent admirers of the Aylesbury with its sturdy well bred appearance and its quiet, docile temperament so essential in putting on flesh at the minimum of cost.

Standard weights are Old drake 9 lbs.; Old duck 8 lbs.; Young drake 8 lbs.; Young duck 7 lbs.

The Pekin Duck

Of the origin of the Pekin Duck, John H. Robinson in his magnificent book *Ducks and Geese for Profit and Pleasure* published some twenty-five years ago says: "Among the young men sent from China to America

A typical Pekin Drake. Courtesy James T. Eiswald, Little Falls, N. J.

about 1870 to complete their education at American Colleges, there was one, Chan Laisun by name who was sent to the Massachusetts Agricultural College. At some time during 1873 he gave an address on Chinese Agriculture before the Hampden Agricultural Society, an abstract of which was published in *Massachusetts Agriculture* for 1873. Neither the place nor the date of the meeting are given in the report; but the place was probably either Westfield, Springfield or Palmer. In describing the live stock of China, Chan Laisun concluded with this statement. 'Chickens of the Cochin tribe and ducks almost as large as geese are common domestic fowls. The former have been introduced into this country but I have not seen the ducks.'

"This was probably the first publicity given the Pekin duck in America. From the fact that it was mentioned as a common fowl, and not given any particular name it appears plain that this Chinaman did not know it as the 'Pekin Duck.' His use of the term 'Cochin' would not imply that Cochins were known by that name in China. They were well known here and he would naturally use the name familiar to his audience. It should be noted as significant that he did not mention the color of the ducks, but only stated that the common ducks of China were almost as large as geese."

At the time this address was made, the first importation of Chinese ducks was either on its way from Shanghai or already in America.

It is possible, though rather improbable, that the first exhibit of them in this country had already been made.

The issue of the *Poultry World* of July 1874 contained the following contribution from George P. Anthony of Westerly, R. I.; "In answer to many inquiries in relation to the Imperial Pekin Duck, unknown in this country or Europe, previous to the Spring of 1873, I give below a brief account of their importation.

"Mr. McGrath of the firm of Fogg and Company engaged in the Japan and China trade. In one of his excursions to China, he first saw them at the city of Pekin and from their large size first thought them a small breed of geese. He succeeded in purchasing a number of eggs and brought them to Shanghai. Placing them under hens, he in due time obtained fifteen ducklings sufficiently mature to ship in charge of Mr. James E. Palmer who was about returning to America. He offered Mr. Palmer one half the birds that he should bring to port alive and the latter accepting the offer took charge of them. Six ducks and three drakes survived the voyage of one hundred and twenty-four days and were landed in New York on the 13th of March 1873. Leaving three ducks and two drakes consigned to parties in New York, to be sent to Mr. McGrath's family (who never received them as they were killed and eaten in the city) Mr. P. took the three remaining ducks and drake to his home at Wequetque [sic] in Stonington, Conn. They soon recovered from the effects of their long voyage and commenced laying the latter part of March, continuing to lay until the last of July. They are very prolific, the three ducks laying about three hundred and twenty-five eggs.

"The ducks are white with yellowish tinge to the underpart of the feathers. Their wings are a little less than medium in length compared with other varieties. They make as little effort to fly as the Asiatic fowls and can as easily be kept in enclosures. Their beaks are yellow, their necks long, their legs short and red.

"The ducks are very large, and uniform in size, weighing at four months old about twelve pounds to the pair. They appear to be very hardy, not minding

severe weather. Water to drink seems to be all that
they require to bring them to a perfect development.
I was more successful in rearing them with only a shallow dish filled to the depth of one inch with water than
those who had the advantage of pond and running
stream."

This interesting account of the first importation of
Pekin ducks to this country seems to establish the fact
that they first arrived here in 1873 and that during the
seventy-three years which have intervened the type
has changed very little from that described by George
P. Anthony of Westerly, R. I. in his letter to the *Poultry World* in 1874.

And in the last seventy-three years no breed of
ducks has had the same amount of publicity, none has
attracted as much attention, and surely none has made
as much money for its growers as the Pekin.

There are several very good reasons for its great and
continued popularity. Any breed of livestock or poultry to have long and continued popularity must have
economic value. It must be attractive in appearance.
It must be free from any definite handicap of size,
shape or color. And it must have energetic and intelligent salesmanship or "showmanship" to put it across
in the first place. And all of these the Pekin has, and
has had from the very first.

The Pekin from the beginning has shown tremendous vitality which is the very foundation of economic
values. It has no handicap of colored feathers nor undue smallness of size, nor is it overlarge. I think it
comes nearest to the ideal business duck for large scale
commercial production of any breed. Perhaps I'm
wrong but I think the seventy-four year record of the
Pekin in America will bear me out.

There is no truth to the statement we hear constantly and that has been made ever since the Pekin

became known in this country that it will do *better* without having access to water in which to swim and bathe itself. James Rankin, noted duck breeder of Massachusetts and one of the first large scale duck growers steadfastly claimed this was so, but many experiments have proven this to be a mistaken idea. Pekins as well as other breeds get along very nicely without access to water in which to swim if they have plenty to drink, but there is no breed of ducks that will do *better* without its natural element.

The Pekin is a large symmetrical, active, white duck, smoother in outline than other large ducks and with a characteristic carriage, head-shape and expression. The typical Pekin bill is rather broad and short for so large a bird. The skull should be broad and full and its round sweep in front forms a sort of "stop" at its juncture with the bill which gives the impression of a concave line from top of skull to tip of bill when viewed at a distance. However, do not get the impression that a concave bill is desirable. Nothing could be further from the fact. The bill should be straight on top or even slightly convex. The apparent concave appearance from the top of the skull to end of bill being caused by the general line from top to tip.

The eye should be rather deep-set in appearance. This is caused by the broad skull and the rather heavy cheeks and which, with the short, stout bill, gives the characteristic "Pekin expression."

The neck should be medium in length, rather thick and well arched, blending nicely into the slanting back and shoulders.

The back is broad for its length; approximately one half its length. The breadth should go back to the well spread tail which aids to this effect. It should be straight and flat and slant straight back from high carried shoulders.

The body should be broad and full with no indication of keel except a little between the legs.

Breast, broad and deep; while not appearing as deep as the Aylesbury and Rouen owing to the absence of keel, it should in fact be fully as deep but owing to the more upright carriage of the Pekin, it is carried higher.

Wings should be rather short and folded smoothly to the body.

Legs, stout and short and set well back on the body. Straight toes connected by the web.

Carriage rather upright but in no way approaching the upright stature of the Runner.

Color: bill a bright orange free from black marks or spots. Eyes, dark, lead-blue. Plumage, white or creamy white. Legs and feet reddish orange.

Standard Weights: Old drake 9 lbs.; Old duck 8 lbs.; Young drake 8 lbs.; Young duck 7 lbs.

The Rouen Duck

The Rouen duck unquestionably is a direct lineal descendant of the wild Mallard through the domesticated Mallard and down through many years of selection by breeders with the object of getting larger, more beautiful and useful ducks.

The color and markings of both breeds are practically identical and it takes but three or four generations of the wild Mallard of today, under domestication and full feeding, to greatly increase its size; and conversely the best of Rouens will degenerate under poor feeding and management well along the road back to the size of the immediate descendant of the Mallards in a very few generations.

In the early days, there was some difference of opinion as to whether or not the city of Rouen could rightly claim the honor of developing the breed but the con-

sensus of opinion among the old writers seems to have
been that the Rouen, or Rhone as it was sometimes
called first in England, came from Rouen or its vicinity
and so that city was entitled to the honor of naming
this grand duck.

Rouens were first introduced into England shortly

A splendid Rouen Duck, Grand Champion
Waterfowl, at Madison Square Garden, 1945.
Exhibited by Harry W. Sterling and Son, Bor-
dentown, N. J.

after 1800 and from all we can learn by reading, they
were at that time about half way in size and massiveness between the Mallard, from which they descended,
and the magnificent Rouens of today. For magnificent
is the word for the modern Rouen.

The rich mahogany brown of the females with the
accurate and distinct dark brown and black pencilling
and the lovely claret, pencilled and shaded greys and
iridescent greens and blues of the drake, with the

great size and massive type of these birds makes them in appearance, perhaps, as impressive as any breed of waterfowl we have.

They first were imported into this country, according to any available records, in the year 1850 by Mr. D. W. Lincoln of Worcester. It is quite certain that had there been any Rouens in America before that time, they would have been shown at the first Boston Poultry Show the year before, or at the New York State Fairs where poultry was just beginning to take its place as a worthwhile exhibit.

Regarding their utilitarian values, there has always been general unanimity of opinion. In 1847 Walter B. Dickson, British authority of the day, in *Domestic Poultry* wrote, "The tame variety (of ducks) most in request is the dark colored Rouen or "Rhone" duck originally from France but now sufficiently common here. These ought to be of the largest size for if they are small, it is probable they are not far removed from the original wild breed, and in that case—will not only be very apt to stray away, but will be less prolific in eggs, though both eggs and the flesh will be higher flavored."

James Main, another writer on poultry of the same period as Dickson and who, it is claimed, had travelled widely in France, writing in his book *A Treatise on Breeding, Rearing and Fattening Poultry,* published in 1847, says of the Rouen "The ducks of the large species are finer in Normandy than in any other Canton in France. The English come often to purchase them in the environs of Rouen to enrich their farm yards and to improve their degenerate or bastard species. Ducks are therefore a trade for the coasting captains of that nation, who, in passing to return home to sell them again to rich land owners."

However, the Reverend Edmund Saul Dixon, one of

the best of the early poultry writers in England, but
not the most authentic, did not seem to have the same
favorable opinion of the Rouen as the French people
or of other Englishmen who wrote of them. Mr. Dixon
describes them as follows:—"Of mottled and pied
sorts, there exist a great variety—black and white,
brown and white, lightly speckled, and many other
mixtures. The Rouen duck of poultry books can
hardly be separated from this miscellaneous rabble
and ought to be permitted to return to its original
obscurity in the multitude. It is wrong to lead people
to pay high prices for them as stock; and we are quite
at a loss to discover in them any unusual merit or de-
scribable peculiarity. They appear to be identical with
the common duck we see everywhere." The Reverend
Dixon seems to be quite in the minority in his poor
opinion of the Rouen which he adheres to in later
editions of his notable and beautifully written book
Useful and Ornamental Poultry first published in 1845.

Wingfield and Johnson in *The Poultry Book* published in 1853 say "We have given precedence to the
Rouen or "Rhone" duck (for both these appellations

A flock of good Standardbred Rouens. Government Experimental
Farm, Ottawa, Canada.

have been conferred on it in confirmation of its French origin) because we believe that under ordinary circumstances it will be found the most profitable variety. Its color, moreover is of great richness. The latter we will at once admit is a point of minor consequence in a bird whose merits must be mainly weighed by its value as an economical habitant of the poultry yard; but where both these recommendations can be combined, there are few persons who would not be desirous of uniting them."

Then, farther along they quote Mr. Hewitt, a famous waterfowl breeder of those days, as follows:— "I am confident that when obtained purely bred, the Rouen is the most profit producing of the duck tribe. They are the most lethargic and consequently the most speedily fed of any; but they lay great numbers of large eggs, the average weight of which would be 3½ ounces—always above three ounces. The color of the egg is a blue green; the shell being considerably thicker than the Aylesbury breed. The flesh is of the highest possible flavor and in first rate specimens the supply is most profuse; for the drake and three ducks belonging to the writer which were so successful for several years at the Midland Poultry Show, when they were weighed by the judges, weighed 26½ pounds, and this too, when simply taken from the pond without any previous preparation. On another occasion when purposely fattened, they attained the almost incredible weight of 34 pounds. I have frequently known the young drakes of only nine weeks old to weigh (when killed) 12 pounds to the pair and in some instances even more than this. As regards their consumption of food I have not found them to eat more than birds of smaller species—they are as hardy as any kind and rarely evince any disposition to wander from the immediate vicinity of the homestead."

Upper—Rouen Drake. Notice the low set, nearly horizontal body, the massive appearance and the arched back. Lower—Pair of Black East India Ducks. (Photographs from the Bureau of Animal Industry, U. S. Department of Agriculture.)

Tegetmeir in the second edition of *The Poultry Book* which is largely a condensation of the first with additions of the writers own opinions, praises the Rouen but seems to favor the Aylesbury. But Harrison Weir, writing in later life says:—"Such was the Rouen duck about fifty years ago. Since then it has risen and fallen both in the fanciers' estimation and in that of the public . . . Yet it is a grand duck and, as grown now, is a still more wonderful bird than it was even thirty years ago. Careful selection, both for color and size has worked wonders and there have been fat drakes that turned the scale at 12 pounds, and ducks over ten pounds; but this large size has been purchased at the expense of fecundity; though still good layers, they are now surpassed by others and much so by common varieties . . . Taking the Rouen for what it is now, it can safely be recommended as the best large, colored duck possibly in existence."

So the evidence of the old writers seems to be that the medium sized Rouens while not the most spectacular and not the ones that win in the shows, are the best for practical business purposes, for which most ducks are kept and that over emphasis of size lowered the economic values then just as it does today. However, to keep up the average weight of the breed and overcome the "drag of the race" which constantly tends to pull the size back toward that of its progenitor the small and lively Mallard, we must have stock birds of the large size to cross back from time to time and so the competition of the show, favoring as it does the large birds, is rendering its own service for the good of the breed just as in other kinds of poultry and livestock where maintaining reasonably large size is desirable.

The modern Rouen is a big, rich looking, impressive duck. In any exhibition of ducks and geese, it seems to me there is more universal admiration of the Rouen than of any other variety.

The head is clean cut with no sign of chubbiness or coarse appearance and joins the neck without throatiness. The bill should be rather long, flattish and not concave but rather straight on top and clean cut and nicely proportioned with the head and neck. A line from the top of skull to end of bill seen in profile gives the concave appearance sometimes attributed to the bill.

The neck is rather long and with no indication of coarseness. On account of the level carriage of the back which is desired in the typical Rouen, the neck rises from it at an acute angle and with a slight and graceful curve and taper as it approaches the head. The eye should be bright and alert in expression and somewhat full.

The back should be broad, nearly level from side to side and slightly convex from front to back.

The body is massive in appearance and deep. It is carried nearly horizontally with a prominent full breast from which the deep keel extends backward horizontally to a point behind the legs where it merges with the underbody. The lower line of the keel should be carried straight with the ground and just clearing it.

The wings should be short and folded nicely against the body and carried rather high, but preferably not meeting over the back. The legs are short and stout and set not too far back on the body; as this necessitates a more upright body carriage than is desired.

While the Rouen probably does not make the same rapid growth at a very early age as the great market duck, the Pekin, still as a general farm duck to be grown under more natural conditions they will probably reach as great weight at from four to five months old; as farm grown ducks are not usually marketed as

"green ducks" and at from four to six months, the
Rouens are past the pinfeather stage, the dark color is
of little, if any, disadvantage and from the standpoint
of attractiveness on farms or in runs where there is no
opportunity for bathing and where the white ducks invariably show dirt and stains, there is a distinct advantage in favor of the darker colors of the Rouens.

As stated before, the color scheme of the Rouen is
not surpassed in any domestic fowl.

The bill of the drake should be a greenish yellow.
The bean a very dark green or bluish black.

The eye dark, brilliant hazel brown and the head
brilliant iridescent green extending down the neck to
the clear white ring that circles the neck at a point
about two inches above the back and which nearly
meets at the back of the neck. From the ring, the neck
and breast from the white collar, or ring, is deep, rich
claret extending to a point well under the breast and
there breaking off sharply where it meets the steel grey
underbody plumage pencilled crosswise with fine black
markings extending back under the stern to the vent,
ending sharply where it meets the deep black extending downward from tail and rump.

The back of the drake, between the shoulders,
should be ashy grey mixed with green, merging into a
lustrous green over the rump, meeting there the tail
coverts of brilliant purplish black.

The main tail feathers are a dark, slatish brown
which in mature specimens show white edging on the
outside feathers. The curly sex-feathers in the tails are
brilliant purplish black as in the coverts.

The wings show a beautiful and intricate color pattern. The main coverts are an ashy grey, shading off
lighter on the bows. The smaller coverts being a
lighter and more delicate grey, finely pencilled with

black with a narrow white bar crossing each feather and ending with a black tip.

The secondaries are a greyish black upon the upper side of the quills and a brilliant iridescent blue on the under side of the quill ending with a black bar tipped with white. This color arrangement results in a brilliant blue band edged on either side with a narrow white bar crossing the wing when folded. The primaries vary from a light to a darker grey and must be free from white on the outer side.

The Rouen duck is a symphony in rich mahogany brown, sharply pencilled with darker brown and brilliant metallic black. The bill is a dull orange with a blue black splash over the top and extending two thirds of the distance from base to tip and reaching over the sides nearly to the edges of the upper mandible. The bean should be purplish black. The head and throat are medium brown with a broad, very dark brown stripe running from the base of bill over the top of the skull and down the back of the neck to the shoulders.

On either side of the head appear two lines of very dark brown, running parallel to each other and reaching back to the eye from the back of mouth.

There should be no suggestion of a white collar on the neck of the duck.

The color of the eye is dark hazel.

The general color of the neck should be a rich mahogany brown.

The wings of the duck have the same wing bar as described for the drake but the wing bow is a very dark brown, laced with mahogany instead of being pencilled like the drake.

The primaries are a dark brown, free from white.

The balance of the body plumage of the duck should be a dark, rich, uniform brown, evenly and sharply pencilled with very dark brown which, as it approaches the tail becomes a brilliant greenish black. White edges on the outside tail feathers are allowable in old birds.

All in all, here is a breed of ducks that will have the attention of the real waterfowl lover just as long as beautiful color and distinctive type are found desirable by man.

The Rouen is of a quiet, docile disposition, putting on flesh economically and fairly quickly when grown. Fair layers of eggs running from bluish green to nearly white, this breed is well fitted for the purposes of the farm, the country estate or the yard of the fancier.

The Standard weights are: Old drake 9 lbs.; Old duck 8 lbs.; Young drake 8 lbs.; Young duck 7 lbs.

I, too, have songs; we also the shepherds call a poet, but I trust them not. For as yet, me thinks, I sing nothing worth of a Varius or a Cinna, but cackle as a goose among melodious swans.

—Virgil, Ecologues.

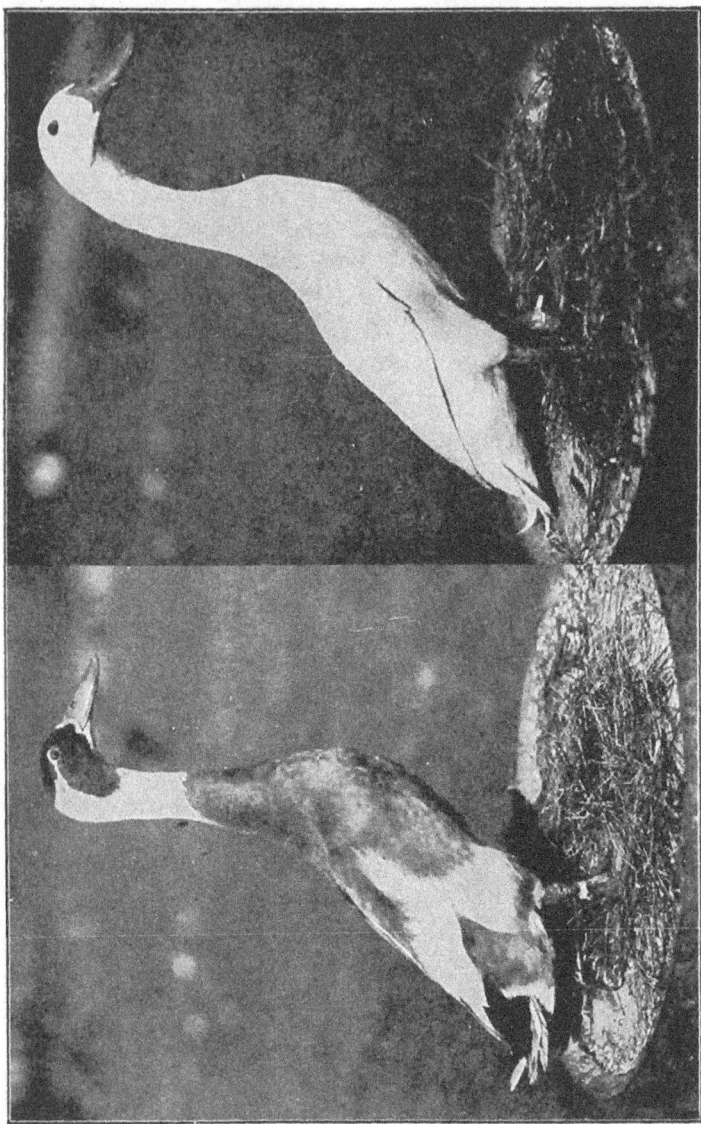

Penciled Runner Drake on left and White Runner Drake on right. (Photographs from the Bureau of Animal Industry, U. S. Department of Agriculture.)

CHAPTER V

Egg Breeds

The Indian Runner Duck

Here is a duck that, in the highly developed show specimens, resembles the exhibition Game chicken. It is tall and erect in carriage, smooth and symmetrical and showing no waste or "lumber"; and is extremely active.

In its accepted mission in life it, however, resembles the Leghorn. It is a heavy, economical and persistent producer of eggs, a non-sitter with a nervous temperament and recognized, like the Leghorn hen, as a very highly specialized laying breed.

Its origin has been from the beginning, a source of sharp and sometimes very bitter arguments. Probably it will never be definitely settled whence its ancestor came or how the Indian Runner was produced.

M. Louis VanderSnickt, the famous continental waterfowl authority, insisted from the beginning of its fame in England where it was first introduced under its present name, that it was nothing more or less than a development of the common farm duck of Belgium which had been bred for centuries by the peasant farmers for heavy and economical production of eggs and that the first Indian Runners seen in England were not in appearance far removed from these common Belgium ducks of the farms and fields.

Realizing the reputation and standing of Mr. VanderSnickt and his vast opportunity for intimate study of the poultry of Western Continental Europe, one

would certainly be inclined to accept his statements as
authentic and so give to Belgium the credit of having
produced the ancestors of the Indian Runner duck and
I believe that has been the general concensus of opinion of those in America who have followed this controversy down through the years.

On the other hand, J. J. Nolan in 1850 in *Ornamental and Domestic Fowl* published in Dublin says:
"The Penguin duck has been imported from Bombay,
and is the domestic duck of that country. They are
more curious than ornamental; they are of a dirty dun
color, with their legs farther behind than the ordinary
duck which obliges them to stand nearly erect when
making an attempt at a quick motion when on land."
And Tegetmeir in *The Poultry Book,* 1877, quotes
Nolan and says: "This very extraordinary looking
duck is characterized by the extreme shortness of the
femur, the thigh or upper bones of the leg; hence their
feet are not brought, as in other ducks, under the middle of the body at an equal distance from the head and
tail, but are placed much farther behind. In consequence of this peculiarity of structure, the duck in
order to bring the center of gravity over the point of
support, is obliged to assume an erect position like
that of the penguin, or other diving birds whose feet
are placed in the same position."

And after the Indian Runner had begun to attract
considerable attention and seemed to be due for a run
of popularity, Mr. Harrison Weir in *Our Poultry* published in 1902 reverted back to the Penguin duck as
follows: "This is one of the most peculiar and remarkable of the duck tribe. Its long narrow head on a thin
neck set on a long, apparently attenuated body, having
an extraordinary upright carriage which last is accounted for by the thighs, legs and shanks being very
short and placed so far back that the bird is obliged to

carry itself very erect to walk or run, which latter it can do with some degree of rapidity. Thus in shape, color and action it much resembles what is now called the Indian Runner duck."

Then Lewis Wright at about the same time in his *Poultry Book* referred to the Indian Runner linking it with "Penguin" ancestry as follows: "The best laying ducks, according to English experience . . . are those now called Indian Runners, which until recent years were little known and received only occasional notice under the name of Penguin ducks."

About this time a Mr. Henry Digby of Grimscar, Huddersfield, wrote in *Feathered Wings*: ". . . during recent years Miss Wilson-Wilson of Kendall, Mr. James Swales, myself and others, have made special effort to trace the origin of this variety and if possible to import fresh birds of the original stock, but in all this direction proved futile until the end of the year 1898 when I succeeded in procuring and importing a trio from a friend in Calcutta which I am pleased to say, has produced a marked improvement in the quality of the stock." Thus the Englishmen seemed to favor the theory of the Indian or Penguin derivation and the Belgian authority, the more logical theory, that they descended from the small heavy laying ducks of Europe so like them in a primitive way. The question will probably never be decided.

But it is significant that if fed as heavily as other ducks and not selected for the long erect cylindrical shape, the Runner tends in a very few generations to go back in type to the farm duck of Belgium, the same type as that fixed by hundreds of years natural breeding by the peasant farmers there, rather than toward the so-called Penguin duck of India; and the evidence of old Mother Nature is hard to refute.

Indian Runners, the name finally universally agreed

on, came into England about the year 1825 and were first exhibited at Kendall in 1897. From that time for the next ten years it gained general popularity, slowly at first but with accelerated momentum till about 1910 to 1914 it reached a real boom and attracted a tremendous amount of interest both from breeders and the press.

Extravagant claims were made for it as a producer of fabulous numbers of eggs and soon like all breeds which have been exploited by exaggerated praise, the flood tide of Indian Runner acclaim began to recede until now they take their place as a really useful egg producing breed on their own merit, without exaggerated praise or undue exploitation.

Indian runners came to this country about 1897 but they did not have the same boom, with its inevitable reaction as in England.

They were admitted to the *Standard of Perfection* at the Boston meeting in 1898 but only the Fawn and White variety. The Pencilled variety and the White being admitted in 1915. There have been other varieties bred spasmodically such as Black and the "Fairy Fawn" and in England the Pied, but these; Fawn and White, Pencilled, and White, only, have been widely bred.

In shape, the Indian Runner, when bred to either our or the English Standard, resembles no other duck. The head is long, blending imperceptibly into the broad flat bill which extends from top of skull in a straight line like a long flat wedge. The eye is set very high. The head and bill, in proportion and correct lines, is most important in giving the Runner its characteristic appearance. The bill should not be convex giving it a roman-nosed appearance nor should it be concave, but as said before should form a straight bold line from the top of skull to tip of bill. Not less im-

portant is the eye setting which should be closer to top
of head than in any other breed; this has in some instances been over emphasized in judging at our shows
but still it does much to give the Runner the keen,

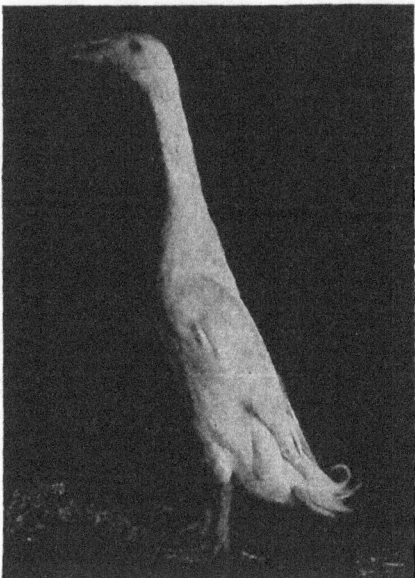

Grand Champion White Indian Runner
Drake, 95th Boston Poultry Show, 1944.
Exhibited by John B. Lightfoot, Orange,
Conn.

sharp expression which adds so much to the character
of the bird.

The head should be carried at practically a right
angle to the long, slim, tapering neck which should be
carried upright and nearly straight, rising from the
slanting back with no angle at its smooth blending
with it. The body is long and cylindrical, feathered
very closely, it should be carried nearly upright with
tail carried straight down in line with the back and

wings folded closely with tips just meeting over the back at rear.

The color of drake and duck in the Fawn and White Indian Runners is the same; an even fawn, free from pencilling and when bred true, with the distinct and regular division of Fawn and White, is very attractive. The head has a line of white extending back from the eye dividing the fawn on top of the head called the cap. From the white neck and throat another narrow white line extends over front of head at base of bill dividing the bill from the head marking.

The bill of the drake is yellow when young; greenish yellow when mature. The bill of the duck, when young, is yellow spotted green and a dull green at maturity. The eyes are dark brown. The lower third of the neck is fawn and the upper part white. The shoulder and upper part of the wings are fawn; primaries and secondaries white extending up over the wing to a point just over the lower part of the body making an inverted V marking on the side of the body. The shoulder and top part of wing fold making a heart shaped marking on the back. Color of the back should be a soft, medium fawn showing an indistinct marking of slightly darker shade.

The tail is fawn; breast is fawn and white about evenly divided between the lower and upper parts.

The body should be a medium shade of fawn, matching the fawn of the breast. Fluff white, except an indefinite area of fawn which runs from base of tail to thigh. Shanks and toes are orange-red.

Male and female color and markings are identical except in the female slightly indistinct darker markings on the fawn are allowed.

The markings of the Pencilled variety are practically the same except instead of being a uniform fawn it should be a slightly darker fawn in the drake with a bronze green head and tail marking and in the duck the uniform, medium fawn should be lightly pencilled with slightly darker fawn.

The color of the White calls for pure white in all plumage sections with leaden blue eyes, yellow bill and bright orange shanks and feet.

The Standard weights for all varieties are Old drake 4½ lbs.; Old duck 4 lbs.; Young drake 4 lbs. and Young duck 3½ lbs.

The Khaki Campbell Duck

Here is a duck on which there seems to be no controversy as to its place of origin. All agree that it was produced in England from the famous utility duck called the Campbell and from the Indian Runner on the other side of the house.

Khaki Campbell Ducks. Courtesy New Jersey State College of Agriculture.

The Campbell was originated in England some fifty years ago and probably contains Rouen blood, as in appearance they resemble the Rouen but lighter in color and not as heavy as the large show-type specimens of this breed. Whether there was other blood used in producing the Campbell or whether it was developed by selection along laying lines from the pure Rouen, will probably always remain a mystery as Mrs. Adale Campbell of Uley in Gloucestershire, England seems to have been a very secretive lady on this point.

With regard to the parentage of the Khaki-Campbell, Lewis Wright in *The Poultry Book* says: "... the Indian Runner has been used in crossing to produce this variety. The Khaki Campbell is of extremely active habits, doing its best on a good range and showing very little desire for swimming—in fact Campbell, we believe, only allows them drinking water.

"At twelve weeks old the ducklings come up to about 4 to 4½ pounds, the laying being about the same average as the other strain (The Campbell). Whatever time of year they are hatched, they are said to commence laying at, or before they are six months old so that by hatching about three lots, very early, medium and late, eggs are obtained every day in the year."

Robinson in *Growing Ducks and Geese for Profit and Pleasure* says: "Campbell Ducks though known in England by that name for about a quarter of a century can hardly be considered a breed. In appearance they seem to be crossed between the Rouen and Indian Runner. They take their name from the originator, a Mrs. Campbell who,—from reports when they were brought out—was averse to saying much about their breeding. Both Wright and Weir in 1902 do describe this duck but Brown in 1906 does not mention it."

Of the Khaki Campbell he says very briefly: "These are a cross of the preceding with the Indian Runner.

The color is described as 'Khaki color' with a darker
lacing on the edge of each feather. The female is of
this color all over. The male has a bronze green head
and tail."

This last is not in accordance with our American
Standard which calls for a brown-bronze head and
warns against the green shade of color.

Now see what a later writer and one apparently very
familiar with this duck has to say. Perhaps we should
allow for a slight, possible and perhaps unintentional
exaggeration of their good points in what Mr. Joseph
Pettipher, a well known British breeder of Khaki
Campbells says of his favorite breed in the *Feathered
World Year Book* for 1920, as he is evidentially an en-
thusiastic breeder and admirer of the Khaki-Campbell.
"The reason why so many people have not seen them
is because of being a purely utility bird, bred to lay
eggs, they are seldom seen at shows. Once or twice in
pre-war days, classes were provided for them at the
Palace etc., but from an exhibitors point of view they
are not a show duck and were never intended to be so.
They were originated by Mrs. Adale Campbell some
twenty years ago specially for egg production, and as
layers they have proved themselves champions. They
vie in that respect with the Runner but have the added
advantage of being better table birds—a consideration
when surplus drakes have to be marketed or killed for
the home table. Personally, I can honestly say that,
after forty years experience of most kind of ducks, I
have never had their equals as layers and today I
think their only rival in this respect is the White Runner, under whatever equal conditions they may be
kept.

"I have had a pen average 286 eggs in a year from the
day they started to lay, and we have heard of higher
isolated instances up from the famous Martock duck

which is vouched to have turned out 360 eggs in 365 days! But there is one point I would especially like to emphasize. I have written it elsewhere, but it cannot be too often repeated, and that is, piracies are rampant. I have seen absolutely crossbred mongrels called Khaki-Campbells, and offered for sale as such and I cannot too strongly warn intending purchasers, either of stock or eggs, to be careful they get the true and original strain. They now have been extensively bred by so many people that it is quite easy to procure fresh blood and so avoid inbreeding; as many breed from different matings each year and follow certain strains, not only to produce the best layers but also to follow distinct lines for the purpose of making out-bred matings. The Khaki-Campbell is Cosmopolitan. It seems to thrive anywhere. My own have free range of field and stream, but I know many people who keep them in more or less confined places with excellent results. Though eggs are the main object these ducks breed true to type as far as it can be considered wise to follow it, though many allow some latitude in that respect, as they are only after the number of eggs. The latter are ALWAYS white-shelled and not over large, but of the most delicate flavor when eaten—another point in their favor. . . . I sometimes meet with people who have become imbued with the notion that some individual ducks of this breed lay all the year round. This of course is not so, generally, though it was in the isolated instance at Martock. If you want eggs all the time you must vary the ages of your ducks. Taken as a general rule the age at which Khakis will commence to lay depends somewhat on the time of year when hatched. Spring hatched birds usually start a little earlier than the later ones. Late March or April hatched will usually lay at from 4½ to 5 months, these brought out at other seasons from 6 to 6½ months.

But though I have frequently had them laying at these early ages I am not particularly keen on it. After all, however good a layer any duck may be, you can but have 'the cat in her skin' and it is better and more conducive to a final good laying record for the bird to get well developed and the productive organs properly established before she starts business."

There is a world of truth in the last part of Mr. Pettipher's statements not only as it applies to his favorite, the Khaki Campbell duck, but to any other variety of ducks and chickens and so I have quoted him rather fully.

The Khaki-Campbell as we breed it in America and to Standard color is a rich looking, sprightly and very attractive duck. Add to this its good table quality and its heavy laying power, which all who have bred it agree on, and it is a fine addition to our breeds of ducks.

It is rather upright in appearance with a long, clean cut head. Its bill is rather long and of medium breadth, nearly straight from the top of head.

The eye is bright and intelligent and set high. The neck is medium in length and with very little curve and carried rather erect. Wings set high and carried closely. Back rather wide and not too long, carried slanting straight from high shoulders to base of tail. Legs of medium length and set well back and well apart.

The head and neck of the drake should be a rich dark brown with bronze sheen: not green. The bill dark green with black bean. Eyes dark brown.

Wings, back, body and breast a rich khaki. Tail coverts and lower back brown-bronze. Underneath tail brown-bronze. Main tail feathers a khaki midway between the body color and tail coverts. Legs and feet rich, dark orange.

Head of the female a rich chocolate brown. Bill and eyes the same as the male. Neck should be:—upper part chocolate brown blending into khaki as it approaches the back.

Wings, back, body and tail an even shade of khaki, the less pencilling the better. Light feathers in wing bar is a defect.

Legs and toes brown, matching as nearly as possible the color of the body plumage.

Standard weights. Old drake 4½ lbs.; Old duck 4½ lbs.; Young drake 4 lbs.; Young duck 4 lbs.

There swims no goose so grey, but soon or late
She finds some honest gander for her mate.
—*Pope, The Wife of Bath; Her Prologue.*

CHAPTER VI

General Purpose Breeds of Ducks

The Blue Swedish Duck

The Blue Swedish is in general shape and character very much like the Pekin except in color. It is somewhat smaller in size. The Pekins usually running larger than the Standard weight and the Swedish somewhat smaller. Its origin, unlike that of its American cousin the Cayuga, seems to be very cosmopolitan. Many different countries claim it but none with very definite testimony and none with which the old writers agree with any unanimity.

Sir Edward Brown in his *Races of Domestic Poultry* published in 1906 says: "All the evidence which we have been able to obtain with regard to this duck is from German sources, as it has held a recognized position in the fatherland for many years and is bred to a considerable extent. It is said that it was introduced in 1835 from Pommern which at that period was a part of Sweden. Hence, the name given to the race. It is not improbable that the same influences which led to the Blue Termonde (A Blue Belgian Breed of Ducks) accounts for the Blue Swedish, and it is quite possible that both own the same progenitors. Upon that question we do not enter as there is no certainty. It is more than likely that these birds came from farther East, and it is suggestive that countries in which blue ducks were found were on the northern coast-line of Europe although blue sports are by no means uncommon in the United Kingdom and in France."

233

Upper—Pair of Mule Ducks. Lower—Pair of Blue Swedish Ducks.
(Photographs from the Bureau of Animal Industry, U. S. Department
of Agriculture.)

Appleyard skips them in his book *Ducks* and only refers to a blue breed as one of the various colors originated by Cook of England as Orpington ducks. Robinson in his *Growing Ducks and Geese for Profit and Pleasure* quotes Theodore Jager who in 1902 wrote as follows: ". . . living witnesses bear testimony that the blue Swedish ducks were bred as far back as 1835, a great many years prior to the introduction of Rouens or Pekins. German poultry shows had classes for Swedish ever since fanciers assembled to compare their birds. Is it not, therefore, somewhat amusing to read sometimes the statements of Belgian poultry writers when they claim the Swedish is a product of the Dutch? More probable is it that the Dutch, recognizing the great market value of the Swedish, imported them in olden time and now seem to think this good duck must be their own. It is human, we admit, to claim a good thing, but for the sake of truth, let the originators have the credit." But Robinson, probably the most thorough and authentic writer on waterfowl this country has ever produced, continues as follows: "Considering the interest in duck growing in the Netherlands dating centuries back and the diversity of the color of the ducks there, it would not be reasonable to suppose that Sweden was the only source for ducks of this color. As to where they originated first, that is quite beyond finding out." Wherever it may have originated and from what breed developed or from what crosses it was produced, the Blue Swedish duck of today is one of our really worthwhile breeds.

The Swedish is a fast growing, hardy, plump bodied bird of just the right size for the market or the average family dinner and while commercially he has the disadvantage of a color other than white, he has the concurrent advantage of being less unsightly and attractive in appearance as the blue color will not show dirt and mud stains like the white; and as the interest in duck raising increases by farmers, back lotters and hobbyists who do not market their ducks when "green" we shall see, I think, this good breed continue to increase in popularity.

The Swedish is more erect in carriage than the Rouen or Aylesbury. In temperament it is alert and lively without being wild and while not noted for high egg production, that may well be increased by breeding and selection just as it has in the so-called egg breeds of ducks.

Like all other blue breeds of poultry, the color is not fixed. Breeding from two blue parents one will be pretty sure to get whites, blacks and blues in the offspring with the blues in the majority. By mating blacks and whites bred from blue parents usually there will appear all blue offspring and these in turn when mated will produce all three colors.

To the hobbyist there is a fascination about breeding these blue breeds that make them extremely interesting.

The bill of the Swedish should be straight and fairly long.

The head fine and oval; the skull narrower than that of the Pekin and the eyes higher and not as deep-set.

The neck is rather long and very slightly arched as in other ducks of rather upright carriage.

The wings are medium in length and folded compactly to the body and carried rather high. The tail is closely folded and carried rather upright.

The body is cobby and plump. It is wide for its length and the breast is full, prominent and deep although there is little, if any, indication of keel on good specimens.

The legs are medium in length but appear longer

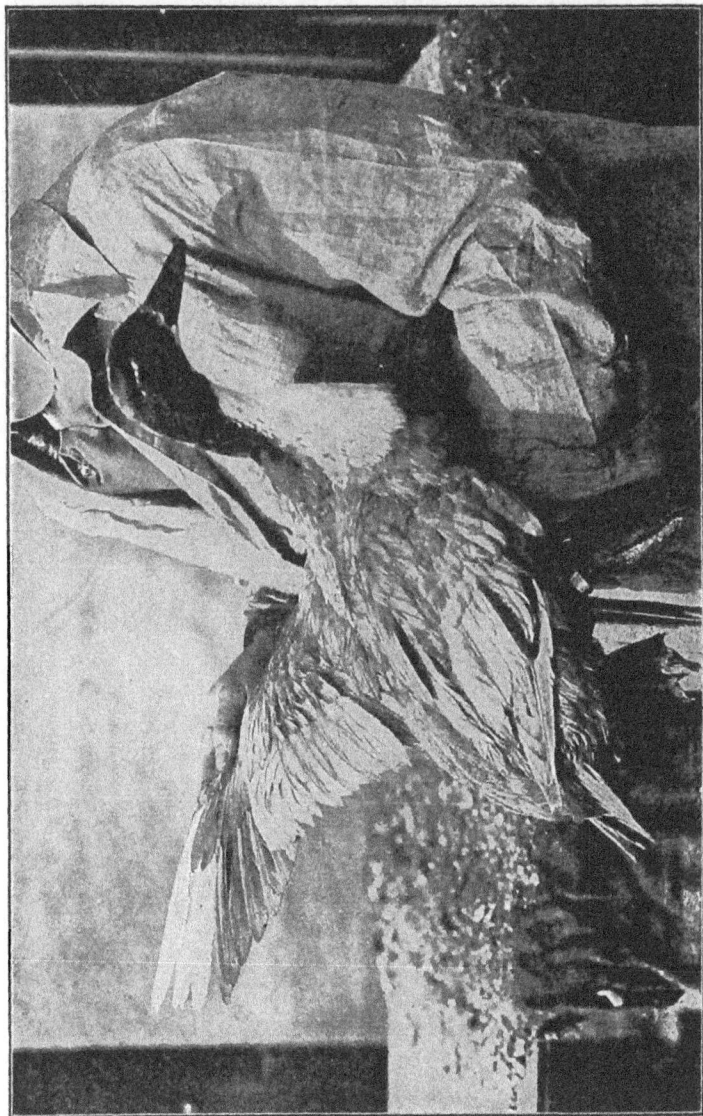

Blue Swedish Duck showing white flight feathers. The Standard calls for only two white flights, but there is a decided tendency as shown here for more flights to be white. (Photograph from the Bureau of Animal Industry, U. S. Department of Agriculture.)

due to the close plumage and the absence of keel, and set back of the body center.

The bill of the drake should be a greenish blue. That of the duck dusky brown with a dark brown "blotch" across the center.

The head of the drake should be very dark blue approaching black, with a brilliant green sheen. That of the duck, a clear steel blue. The body color of both drake and duck should be uniform steel blue except for that of the breast on which there should be a heart shaped spot of pure white about two inches wide by three long extending from a point just below the throat downward, also the two outside primary or flight feathers of the wing should be pure white.

Herein is contained one of the fascinating features of breeding Swedish ducks. The amount of white is not a fixed character in breast or wing and nothing adds so much to the characteristic appearance of this duck as to have the white "bib" just the right size and the white flights of just the right number. More than half white in plumage disqualifies the Swedish and if the bib and white flights are lacking, the feature that adds so much to their smart appearance, you have a very plain and somber appearing duck.

It is admitted that there is no market value in these little embellishments of color markings but there is no economic value to a good looking necktie or those nylons so necessary to the happiness of the ladies; and "man does not live by bread alone."

If one is looking for a breed of ducks that will be economically satisfactory in small scale duck raising or to breed as a hobby, that from the fact that there is always a market for good plump bodied ducks will be self-supporting or a little better, the Blue Swedish will fill the bill.

Weights are—Old drake 8 lbs.; Old duck 7 lbs.;
Young drake 6½ lbs.; Young duck 5½ lbs.

The Buff Duck

Had the Buff Duck ever become widely popular it
would have been by virtue of its own sheer economic
merit and good looks as no one has ever done much of
anything in the way of helping it along the road to
popularity; in this country at least. Even the American Poultry Association fails to give it the dozen or so
grudging words in the introduction to the duck division of the *American Standard of Perfection,* custom-
arily vouchsafed to varieties recognized therein by
that august body.

It skips any word of introduction or of the place it
best fills among the varieties of ducks. Just the head-
ing "Buff Duck" and a brief and rather perfunctory
description of size, color and shape and lets it go at
that. In England it is known as the Buff Orpington
duck after the Orpington farm of William Cook, credited with having originated it in Britain. Over here
we do not even dignify it with a descriptive adjective;
it's just the "Buff." The sponsors of this fine duck and
the committee on Standards which recommended its
admission to the *Standard of Perfection* must have
had little imagination and even less interest when they
recommended for the name just simply "Buff." It's
like naming a breed of chickens "White," a breed of
cattle "Yellow" or a family of horses "Bay." Small
wonder the name has never attracted the attention of
the public and that the breed has never forged ahead
and flourished. It has been recognized by the American Poultry Association in the *Standard of Perfection*
for some thirty years and still, ask the average poultryman or even the average poultry judge whether or

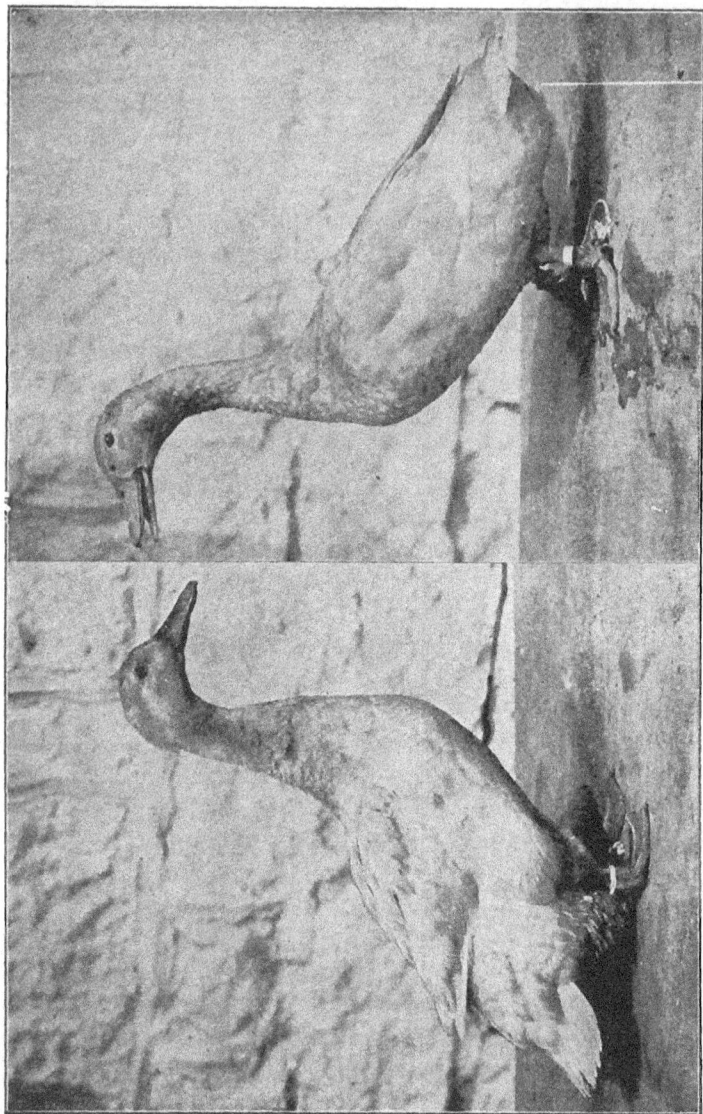

Pair of Buff Ducks—Drake on the right. (Photographs from the Bureau of Animal Industry, U. S. Department of Agriculture.)

no the Buff Duck is in the Standard and the chances
are that he could not tell.

And yet, handicapped as it has been by the most
drab and colorless name of any Standard breed either
of land or waterfowl, this fine, useful duck has gradually improved in size, type, uniformity and color until
today it can take its place in usefulness and good looks
with any of the general purpose ducks. But the name
is a serious handicap. "Buff Duck!" The American
Waterfowl Association should do something about it.

VanderSnickt, the great poultry authority of Belgium and western continental Europe wrote of solid
buff colored ducks among the varied colors in the flocks
of farm ducks of Belgium from which it is probable the
Indian Runners were really developed by breeding and
selection. Is it too much to think that this might be
another descendant of these hardy and prolific ducks
which for so many years had been developed and improved generation after generation purely for economic
purposes by the thrifty peasant farmers of the Belgian
lowlands?

And how much more appeal would there have been
to the name had their sponsors in the early day called
them British Buffs, Belgian Buffs or Buff Continen-
tals. And who can say that had they been thus favored by the men responsible for naming them and
backed as they are by their undeniable good looks and
economic values, they might not be today at or near
the front in popular favor in the whole list of ducks in
America instead of among the least known varieties?

The first Buffs came to this country from England
in 1908 when Wm. Cook, the great British poultry
promotor showed a pair at Madison Square Garden.
Mr. Cook claimed to have made them by using Aylesbury, with Indian Runners, Cayugas and Rouens in
the melting pot and then by selection, fixing the type

and color he wanted which he named Buff Orpington ducks, probably after the Buff Orpington chickens which he originated on his farm at Orpington in Kent which had made such a furore sometime before.

His account of the making of this breed was vague and general and it seems that such a very practical and skillful breeder as Mr. Cook was would have taken the much easier and shorter route to his goal, the Buff Orpington Duck, by simply taking the Buff colored Belgian ducks, a fine useful breed in itself, and crossing with the great English meat duck, the Aylesbury and then by breeding and selection achieving the same end with much less trouble and in much less time.

Appleyard, in his English book *Ducks* says of the Buff Orpington, "From a utility point of view the Buff is of course the best, as it has been bred for eggs, and many ducks of this breed have put up good records at the laying tests. The Buff required is quite a light shade and the underfluff and any 'stubs' in ducklings do not show so much as in the Blues and Blacks (Orpingtons). When killed for table purposes at 11 to 12 weeks old, Buff Orpington drakes are really decent table ducklings, both in color and texture of skin and flesh if properly fattened and finished off."

The British Standard gives it more consideration in its description than does ours, saying "The breed is essentially a triple breed. It combines beauty of form and color with good table quality and profitable egg production. It is contrary to the best interests of the breed and at variance with the correct interpretation of this Standard to breed for any one of these three qualities at the expense of the others." Very well said indeed; and it might be reiterated many times regarding many other breeds of both ducks and chickens as well.

Robinson says of the Buff: "A good bird of this color

makes a very striking and handsome duck, but few of them are seen. Most of the Buff ducks on exhibition discourage interest in Buff ducks."

This was written some twenty-five years ago and is just as true today but they have improved in uniformity and in color and type so that they are really most impressive when put down right, at the shows.

Really there is no better general purpose duck I think. They are splendid layers of large eggs. They will not equal the best of the Runners or Khaki Campbells in egg production as a rule but they have something those breeds lack in a first rate table carcass with more size and with a plumage color so light they do not show stubs badly when plucked; as Appleyard points out in his description of their good qualities.

All this good duck needs is for some one to take it up and develop it still more, both for looks and the useful qualities and then let the public know more about the high degree of both contained in the breed, for it to advance far in public favor.

In size, the Buff is about medium; slightly larger than the Blue Swedish, making it ideal for the general farm duck. It is active and a good forager and strong and robust.

The head is fine and breedy looking. The bill is rather long and straight in line from skull: yellow in the drake and dark orange in the duck, with a black bean. Eyes are brown with a blue pupil and the legs and feet are short and strong, set well apart; orange in color. The neck is fairly long and slightly curved and the wings are short and carried close to the body and closely folded. The back is broad and not too long. The tail is small and slanted up from the back line with the sex feathers of the male hard and well curled. The breast is broad and not too deep, carried well forward and rather low. The body is broad and medium in length but rather deep and carried approaching the horizontal.

The color of the plumage should be an even fawn-buff except the upper neck and head of the drake which should be a rich seal-brown. Pencilling in any section is undesirable and the less the better. Evenness of color is more important than the exact shade.

Weights are Old drake 8 lbs.; Old duck 7 lbs.; Young drake 7 lbs.; Young duck 6 lbs.

The Cayuga Duck

The Cayuga duck is as distinctly American as the Plymouth Rock chicken, the Morgan horse or the Concord grape. All writers seem to agree with the tradition that it descended from the wild black-duck that was first domesticated and crossed with the common duck; or bred pure from a wild foundation in the vicinity of Cayuga Lake in Cayuga County, New York.

Probably the first notice of the Cayuga which associated it with the Cayuga Lake district, was printed in the Albany *Cultivator* published in December 1851 which said: "VARIETY OF DUCKS—We saw at Colonel Sherwood's in Auburn last summer, a singular variety of ducks, and on inquiry were told that they were obtained from Mr. John S. Clark of Throopsville, Cayuga County. We were so much interested in their appearance, especially for their resemblance to the wild black-duck (anas obscura) that we wrote to Mr. Clark to learn their history. In reply he said: 'The ducks you inquire about have been bred distinct from any other variety at least twenty years. We obtained them some ten years since in Orange County and were told that they were originally descended from the wild black-duck, and from the great resemblance, I have no

Upper—Cayuga Duck. Lower—Cayuga Drake. (Photographs from the Bureau of Animal Industry, U. S. Department of Agriculture.)

doubt the statement is true but cannot affirm this as a certainty.

" 'The characteristics of this variety are, nearly a uniform color (a little darker than the wild black-duck) good size, attaining the weight of eight pounds dressed at four months old, very quiet and very prolific, one duck laying from 150 to 200 eggs in a season with proper care.' "

The next printed account of this duck and the first I think where it was called the Cayuga Black duck, was an article by Caleb N. Bement, author of the *American Poulterers Companion,* written for the *Albany Cultivator* in 1863. In this article he seems to bear out the statements of Colonel Sherwood and Mr. Clark printed in the same paper 12 years previously. Mr. Bement writes as follows: "This bird derives its name from the lake on which it is first supposed to have been discovered. But of its origin, like that of the domestic fowl, little is known. It is very natural to inquire whence so remarkable and valuable a bird was originally obtained: but the conclusion seems to be that it results from the intermixture of the wild black-duck (anas obscura) not uncommon in our lakes and rivers; this appears to be the popular opinion at the present time and if we are limited to any one of the wild breeds of this genus, now known to us, it is to the wild black-duck, in our humble opinion, the honor should go.

"This species has, as we are informed, been domesticated in several places, and was quite common some fifty years ago in barnyards in the vicinity of Boston and elsewhere. Says Dr. Bachman in a note addressed to Mr. Audubon: 'In the year 1812 I saw in Duchess County, in the State of New York, at the house of a miller, a fine flock of ducks to the number of at least thirty, which from their peculiar appearance struck me

as different from anything I had before seen among
the varieties of the tame duck. On inquiry, I was informed that three years before, a pair of these ducks
had been captured in the mill pond. They were kept
in the poultry yard, and—it is said—were easily
tamed. One joint of the wing was taken off to prevent
their flying away. In the following spring they were
allowed to go into the pond, and they returned to the
house to be fed. They built their nest on the edge of
the pond and reared large broods. The family of the
miller used them occasionally as food and considered
them equal in flavor to the common duck. The old
males were more beautiful than any I have examined
since, and yet domestication has produced no variation in their plumage.' "

Thus, all early writers to whose writings I have been
able to have access, including the two quoted, agree
that here is a good, medium sized duck; a pure American product, handsome, of good table quality and excellent layers.

Why has the Cayuga never become more popular in
this country? Indeed, it has had vogues: Periods when
it seemed that it was really on the way to wide-spread
popularity. James Rankin, of South Easton, Mass.,
pioneer large scale duck grower who did so much to
establish commercial duck raising in this country after
trying the "puddle ducks," the Aylesbury and the
Rouen says of the Cayuga in 1889: "We found the
Cayuga was a more precocious bird than the Rouen
and were better layers. The eggs were more fertile.
They were much hardier, and, as a consequence there
was less mortality among the young. But they were
rather small in size, dressing only seven to nine pounds
per pair. The Rouens were nearly four pounds heavier
per pair but had their disadvantages. They were not
so productive in eggs and these did not give the same

percentage of hatch, while the mortality among the little ones was much greater . . ."

From this, it appears possible that had not the Pekin appeared on the scene at about this time and caught the attention of this premier promoter and popularizer of the commercial duck business or had the Cayuga been white in color it might have had an entirely different history: and this last is probably the main reason for the Cayuga never having gained widespread popularity as a market duck.

Any breed of poultry, ducks, geese or chickens, must have as few handicaps as possible from a commercial viewpoint to attain and maintain wide popularity and the black plumage, however beautiful, is a distinct disadvantage from a marketing standpoint.

Today we have the Cayuga occupying the position of an old favorite which has been crowded one side to a certain extent by the newer and more publicized ducks such as the Buff, Khaki Campbell and Blue Swedish, all in about the same class of dual purpose ducks particularly suitable for the farm flock or that of the subsistence or part-time farmer or the breeder of ducks as a hobby. A breed with intrinsic value such as this, however, will never die. It may never become widely popular but will always have its friends and admirers.

The Cayuga in general appearance resembles the Rouen in shape more than any other breed. On a smaller scale and of a different color, it has the general outline of the Rouen but on a slightly more refined scale and with less development of keel.

The head is finely chiseled and not coarse with a brilliant, dark eye not placed too high. The bill rather long and broad and not concave. The neck is fine and nicely curved, rising from the back at a rather acute angle.

The carriage should be level with the legs placed not too far back as this makes the duck stand too erect in order to preserve balance.

The back is rather long and fairly broad and should be level; neither concave nor convex and the breadth carried well back.

The breast should be full and carried low and the body deep for its width.

The wings medium in length and carried rather high, nicely folded to the body and the points not meeting over the back. The keel should be well down in front, tapering back to where it merges with the underbody in the rear of legs.

The tail should be folded rather closely and not as erect as in the Pekin, Rouen or Aylesbury.

In color, the Cayuga should be a brilliant, greenish black throughout. The words "should be" are used advisedly as there are very few Cayugas that live up to this color description. The drake is more brilliant than the duck and a brownish cast on the feathers is not a disqualification but a defect differing in seriousness with the degree of brown. White in the plumage in any part is a Standard disqualification.

The bill should be black. The eyes dark brown and the legs and feet a very dark slate.

The Standard weights are: Old drake 8 lbs.; Old duck 7 lbs.; Young drake 7 lbs.; Young duck 6 lbs.

The Crested White Duck

The Crested White Duck is an excellent example of a bird that is unusual and attractive enough in appearance to make it desirable as an ornament to farm or home grounds, pond, pool or stream and to attract the attention of friends and visitors, to make an interest-

Upper—Crested White Drake. Lower—Young White Muscovy Duck showing black on top of the head. Foreign color in the plumage of a White Muscovy is a Standard Disqualification. (Photographs from the Bureau of Animal Industry, U. S. Department of Agriculture.)

ing and zestful hobby and finally of enough economic worth to more than pay its way in eggs and meat.

I have never known anyone to express anything but admiration for a flock of White Crested Ducks or even single individuals. With their rather short cobby bodies, active gait, pure white plumage, blue eyes, bright yellow legs and feet, plus large round crests like ladies powder puffs on top of their heads, they make an unusually attractive sight either on land or water. Of their origin there is no authentic data. "Ducks with top-knots" have been mentioned by the very earliest writers but casually and as isolated instances and of various colors. The earliest work in which we find more than a very brief mention of the Crested Duck is in *The American Poultry Yard,* by D. J. Browne written 1849 where we read: "This is a beautiful and ornamental variety. They are of all colors, having in fact no other common features. We have had them pure white, black, and mixed black and white, with large turbans, or top-knots. . . . The white are considered the most beautiful as they have yellow bills and legs. We are not advised of the origin of this variety. Mr. Brent regards them as probable descendants of the Australian tufted ducks, of which more than one variety is said to exist in that country. Many may have been thus bred; but the accidental appearance of the tuft on the goose, of which bird we have no rumor of any variety thus uniformly decorated, suggests the probability of the duck having obtained its head-dress in a similar accidental manner. The topknot of the latter, however, is in general, proportionately larger and more spherical than of any geese we have ever yet seen, sometimes though placed on the back of the head, even rivalling in form that of the Polish hen. Main speaks of the 'red Crested Duck' from New Zealand but which is not very common there. A red crest grows on its head; a very glossy black grey is predominant on the back, and a deep greyish soot-color on the belly; the bill and legs are lead color, the irides golden.

"Latham also speaks of the Crested duck, and says 'This inhabitant of the extremity of America is of the size of the wild duck but is much longer, for it measures twenty-five inches in length; a tuft adorns its head; a straw yellow, mixed with rusty colored spots, is spread over the throat and front of neck; the wing speculum, blue beneath, edged with white; irides red and the rest of the body ashy grey.' It is a question, therefore not easily answered whether the domestic Crested duck has been produced from a cross of our wood duck, or from the above mentioned variety. If it sprung from either, its size would indicate that the one mentioned by Latham would appear the most likely to produce them. Very fine specimens of the Crested Ducks have been exhibited at the different meetings of Poultry and Agricultural exhibitions."

Jager, in Weir's *Poultry Book*, American Edition edited by Johnson and Brown, 1915 says: "THE CRESTED WHITE DUCK OR TOP-KNOTTED DUCK—This specimen is of very ancient origin and doubtless was a freak of nature in the first place. They may be seen in many of the old Dutch paintings, some almost black, while others are pure white with splashed markings on a white ground being very common. Some of the best I have seen were in my own possession, having very large top-knots—and drakes especially—well covering the head. These top-knots were of long, full sized rounded feathers growing from a considerable fatty substance excresence on the top of the head, though somewhat towards the back. In not many cases this was loose, almost pendulous in character, while in others it firmly adhered to the skull. In no instance was the bone in any way altered from that of the ordinary duck. The eyes were somewhat larger, and the bill rather shorter than that of the Mallard also broader and more rounded at the tip. They were medium thick in girth, lively and quick in their movements, the ducklings especially being excellent fly catchers. They were plentiful layers of fair-sized eggs of a light, bluish green. My birds were very handsome in color; the ducks mostly grey with white about the head, neck and breasts; some being black with white fronts: the drakes being of the bright Mallard hues splashed with white . . . To the detriment of the variety it must be said that a large crest evenly carried, generally heads the winner, even if other more highly counted body points were lacking."

This last is so many times true of other breeds, not only in the duck family but other fowls where some one spectacular feature, of no economic value actually, but difficult to attain, is given undue importance by breeders and judges alike and the more important sections of the breed are let deteriorate until in some cases the breed has been ruined and become practically extinct; for no breed of fowl can long endure, that is deficient in the economic or useful values.

I am inclined to agree with Jager on the origin of the Crested White duck. That the crest is the result of a mutation or sport in the first place and by long years of selection by man the breed's most striking feature, has come to be a fixed characteristic.

Flocks of this breed will produce practically one hundred percent crested young, altho the shape and size of the crest is not fixed, and this makes one of the fascinations of breeding Crested Whites.

The head of the Crested White should be cobby and at the same time refined. Not nearly as rugged as the Pekin, with a more prominent eye of blue and a bright yellow bill with no black. The crest should be large and round and perfectly balanced and set well back on the head. The neck should be rather short and slightly curved, rising from the back at a not too acute angle.

The back should be broad and rather short and slightly convex. The shoulder high, slanting back to the tail which should be carried rather upright. The wings should be well folded and carried close to the sides and well up on the body.

The legs should be set somewhat back of the middle of the body to allow the bird a rather upright and alert position when standing or walking. Color of the legs and feet should be light orange.

Type and size in this breed is fully as important as in other varieties of ducks but has been pretty generally neglected due to over emphasis on crest, the spectacular feature of the breed. Many strains have been crossed with the Pekin to increase size which has injured the natural character of the Crested White. Breed for a cobby, medium sized,. general purpose duck, active and attractive and do not let the crest receive all of your attention.

Weights are Old drake 7 lbs.; Old duck 6 lbs.; Young drake 6 lbs.; Young duck 5 lbs.

The Muscovy Duck or *Pato*

This is unquestionably one of our most useful and valuable breeds of domestic waterfowl. Of its origin there is no doubt or question. It is the direct descendant of the wild Brazilian duck of South America and has been little changed by domestication from its original appearance and characteristics.

The Muscovy is a rugged individualist of the first degree. It retains its unique and intriguing personality no matter under what conditions it is kept. It is a self-

Upper—Colored Muscovy Drake. Notice the partly erect crest feather on top of the head. Lower—White Muscovy Drake. Notice the long, horizontal body and the rough or carunculated face. (Photographs from the Bureau of Animal Industry, U. S. Department of Agriculture.)

respecting, self-assured and well nigh self-supporting bird if kept under conditions where it can garner its own living from grass, seeds, worms and insects. It will make its own nest and hatch and raise its own young with no assistance from man but protection from predators and enough food to supplement what it can glean from field and stream where these are available.

It is not uncommon for a Muscovy duck to steal her nest and in five weeks come forth with from twelve to eighteen young ones and to raise one hundred percent of these with no assistance from human hands except the necessary feed thrown out to supplement the gleanings from the fields—and access to drinking water. She has all the intense mother instinct of the wild bird and will pretty well defend her young from dogs or prowling cats, putting up a wicked and effective defense when necessary.

Of the origin of the Muscovy, there is complete unanimity of opinion, but there agreement pretty well stops regarding this quaint and curious bird. Even its species is not agreed on between writers from the very earliest days. Willoughby called it a goose and this has been the opinion of many: but with very little reason it seems to me. The principal argument is that it must be of different species from the duck as it will not successfully cross with the common duck and the period of incubation takes a week longer. But why this makes it a goose I fail to see as it also will not cross with the goose and there is the same difference in the incubation period. They say it cannot quack like other ducks but neither can it honk like a goose; it being the only domestic waterfowl that will not disturb the neighbors by either of these sounds, it being practically mute.

These arguments remind me of the Maine, Rhode

Island Red fancier who was shown a flock of the so-called "utility" Reds some years ago and remarked "well they call them Reds and I suppose they must be Reds because if they aren't Reds what in hell are they?" So if a Muscovy isn't a duck what is it?

Robinson in *Raising Ducks and Geese for Profit and Pleasure,* now long out of print, contributes the following on the subject: ". . . Among naturalists and poultry keepers in Europe and North America, the nondescript character of the bird led to some difficulty in classifying it. Not properly a duck it is still so like a duck in general appearance that most people call it a duck at sight. Then when closer inspection raises doubt as to the correctness of the identification, the impulse is to assign it to the goose family. Here again there is difficulty, for it is even less like a goose than a duck.

"But as all domestic waterfowl were either ducks or geese, the anomalous creature had to go—according to European precedents—into one group or the other; so, by common consent, naturalists and poultry keepers have called it a duck. The naturalists called it Anas moschata, the musk duck, giving it the name on account of an alleged slight musky odor. Poultry keepers gave it a great variety of names, mostly places from which it was erroneously supposed to have come. It was called Barbary Duck, Guinea Duck, Indian Duck, Turkish Duck, Cairon Duck—as well as Musk Duck—and also by corruption of the last name it has been called Muscovite Duck and Muscovy Duck.

"In the land of its nativity, there being no doubt whatever as to its origin, and no doubt that it was neither duck nor goose, they have given it a name of its own. There they call it 'pato.' The chapter on Waterfowl in Dr. F. F. DeMoraes' *Avicultura* has the title 'Patos,' Marrecos, Gansos, e Cysnes—Patos,

Ducks, Geese and Swans. Further the chapter begins with a description of the 'National Pato,' and an assertion of a title to a name of its own, and, as many poultrymen in the United States know how zealous he has been in working for the development of the poultry industry in Brazil it seems to us eminently proper in a book of this character to give the facts in the case and to recognize the rights of the pato to the extent of calling it by that one of its numerous names which will convey the most accurate information as to its origin, namely—the Brazilian Duck." And with this viewpoint the present writer wholeheartedly concurs.

The pato is manifestly not a true duck in the sense that it descended from the same ancestor as most other domestic ducks apparently have; i.e. the Wild Mallard; their varied characteristics and colors having changed through the development of chance mutations and by selection, generation after generation, for special characteristics to the point where now they appear as distinct breeds.

The proof of this is the difference in the length of incubation of the Muscovy's eggs; its almost entire absence of voice which is confined to a sort of feeble, hoarse and poor imitation of a quack; and then only when very much disturbed or alarmed, and this confined to the female. The character of its tail carriage which unlike all descendants of the Mallard, is carried straight out, horizontal with the body and last and finally, I offer the fact that if, as occasionally happens, it crosses with one of the other varieties of domestic ducks the resulting offspring are hybrids or "mules" as has been proven in a number of instances on the writer's own farm.

While it would probably be impossible to effect the recognition by scientists of the Muscovy, Pato or Brazilian-duck as a distinct species it might well be worthwhile for the American Waterfowl Association to request of the American Poultry Association the change of name for this breed from the perfectly meaningless title of Muscovy to the logical and informative name of BRAZILIAN. It would be a sensible move and a fine gesture in the present movement toward Pan-American friendship. As in the case of the Indian Game fowl which was changed to Cornish by edict of the A.P.A. it would take but a comparatively short time to effect the change.

However, disregarding the name, we have in this duck one of the most useful breeds of domestic fowls; and one of the least appreciated and understood; and most maligned down through the years.

Moubray dismissed it with a half dozen words. The Reverend Edmund Saul Dixon writing in his *Domestic and Ornamental Poultry* in 1848 is perhaps farther from the facts than any other of the old writers. He says: "A pair that were given me by a friend came home very dirty, and as at the end of three or four days they still had not washed themselves from their coating of filth, they were driven toward the water to bathe and be clean. In vain—they flew back over our heads to the poultry yard; and we were obliged to catch them, put them back into the pond with our hands and give them a ducking, whether they would or not.

"Another curious fact is that the great part of the feathers of the musk duck do not resist the wet so well as those of other water birds; but the quill feathers particularly, and those of the tail become so soaked and matted in a very few minutes, like those of a hen or turkey; so that if compelled to make any long voyage at a certain distance, they would sink and be suffocated . . . The female of the musk duck has considerable power of flight and is easy and self-assured in the use of its wings. It is fond of perching on the top of barns, walls, etc. Its feet appear, by their form, to be more adopted to such purposes than those of other ducks. If allowed to spend the night in the hen house, the female will generally go to roost by the side of the hens, but the drake is too heavy to mount thither with ease. Its claws are sharp and long; and it approaches the tribe of Rasores or Scratchers in an unscientific sense, being almost as dangerous to handle incautiously as an ill-tempered cat; and will occasionally adopt a still more offensive and indescribable means of annoyance. The voice of the male is a hoarse, asthmatical sigh. Its habits are gross, obscene, dirty, indolent, selfish. It manifests little affection to its female partner and none toward its offspring. The possession of three or four mates suits it, and them, better than to be confined to a single one. It bullies other fowls sometimes by pulling their feathers, but more frequently by following them close and repeatedly thrusting its face in their way, with an offensive and satyr like expression of countenance. I have not observed it to sift the water as it were, with its bill to extract minute insects and worms as is the custom of other ducks. It is a great devourer of slugs. A nest of half fledged birds of any moderate size would soon be swallowed . . ."

After the foregoing tirade, of exaggerations and misstatements mostly, he does however admit, "It is strange that a dish should now be so much out of fashion as scarcely to be seen or tasted, which, under the name of Guinea Duck, graced every feast 150 years ago, and added dignity to every table at which it was produced."

The truth of the matter is that the Muscovy is one of our most valuable ducks. It is quiet, easily raised, a good layer but not in the class with the Runners and others that have bred for extreme high production. It is practically disease free. Hardy to the highest degree. If given free range, it will probably glean more of its own living than any other breed as it is almost in a class with the goose as a grazer.

The flesh is superior to any other breed and while the Muscovy does not feather as quickly as the Pekin, thus making it less suitable for the "green duck" market, it is a highly desirable market duck for selling later in the season.

In raising Muscovys for market the disparity in size which is greater than in any other breed of waterfowl is really no disadvantage as, though there is a great difference, there are really but two sizes. The drakes are very uniform in size as are the ducks, thus giving two weights, each with very little variation, which makes them attractive in packing if the drakes are packed together and the ducks by themselves.

The Muscovy dresses very attractively, the skin being yellow and the body plump. The flesh is of higher quality than the average duck and for home consumption or local markets is more desirable.

Many strains are prolific layers, often laying through the colder winter months when given good fare and feed. Their eggs are of a creamy white in color and are exceptionally tough shelled and, therefore, will stand much rougher treatment than will the more common duck eggs.

The Standard lists but two varieties of Muscovys; the Colored and the White but a third variety, the Blue, seems to be gaining wide recognition among Muscovy breeders.

The Muscovy has been a fairly common farm and fancier's duck in America for a hundred years or more. At the first Boston Poultry Show in 1849, W. F. Churchill of Roxbury, B. W. Blach of Dedham and

Jonah C. Ellis of Walpole exhibited Muscovy ducks and from that time at least they have been fairly well known in this Country. They have never been widely raised for market, the common objection being the extra week required for incubation, the disparity in the sizes of male and female and at first the color. This last objection has been met by the advent of the White Muscovy, probably a sport from the dark variety originally.

The head of the Muscovy drake is large and coarse, that of the duck more moderate in size and finer in form. The heads of both are adorned with a sort of crest of long feathers which normally are not observable, owing to their being carried closely folded over the back part of the skull but which are quickly elevated when the bird becomes disturbed or alarmed. The sides of the head in both sexes should be covered with bright red caruncles similar to those on the head and neck of a turkey. Those of the drake are considerably larger and better developed than on the female, but in either sex the larger these caruncles and the rougher the face the better the specimen is considered in this particular. These caruncles are one of the features of the breed and as a consequence, smooth faces are undesirable. A feature which distinguishes the drake from the duck is a fleshy knob at the base of the upper bill. This knob when well developed adds to the quality of the specimen.

The bill is rather small for so large a duck, being proportionately shorter and narrower than in other larger breeds. The top line of the bill is very slightly concave, the front part of the skull carrying out this effect. The bill is set onto the head in line with the top of the eye, which is located somewhat lower in the skull than is characteristic of other domestic ducks.

The neck is medium in length and carried well forward, the head not being carried especially high, so, despite the horizontal carriage, there is no such acute angle at the juncture of the neck and shoulders as appears in the horizontally stationed Rouen or Aylesbury.

The back is long; a long, fairly well spread horizontal tail increasing its apparent length considerably, with the result that the distance from the back of the neck to the tip of the tail often exceeds the width across the shoulders as much as two and one half times in the more typical specimens. The Standard does not specify whether the back should be straight, arched or concave, but a large majority of the winners at the leading shows reveal a slightly arched contour over the shoulders, where is also the point of greatest width.

The breast is broad and full but not prominent, and the body is long and wide in the same ratio as the back, but possesses only fair depth. Keel is not desirable in the Muscovy nor is excessive abdominal development, although old females are sometimes inclined to appear rather full in the posterior.

The wings are long and strong, enabling this breed to fly readily. It is therefore necessary to clip one wing to confine them and for this reason, such practice does not disqualify the entry in the showroom. The legs are short and located in the center of the body, resulting in the horizontal carriage characteristic of all Muscovys. The plumage is hard and close fitting and more resembles the feathers of land fowls than does that of other ducks. The sex feathers which distinguish the males of other breeds of ducks are absent on the Muscovy.

The Colored Muscovy

The bill of the Colored variety is a light horn or flesh color and the head black and white while the face is covered with bright red caruncles. The neck may also possess a very little white but the remainder of the plumage, except for a small patch of white upon each wing, should be a glossy, green-black. The white sections of the plumage should be as restricted and as sharply defined as possible, although there is a tendency for real dark individuals to come with the objectionable black faces. Any inclination to grey in the plumage is a fault to be avoided for the reason there is a tendency for individuals of this variety to moult in more white feathers as they age, and it has been found that the grey plumage birds fail to hold their color as well as the more sharply marked ones. White covering more than one half of the entire plumage in the colored variety is ground for disqualification, so the proportion of white must be kept well restricted.

The shanks and feet may vary from a dark lead color to a light yellow, the latter color being less liable to be associated with the black or gypsy face and as a consequence should be preferred.

The White Muscovy

The White Muscovy is white in all sections with flesh colored bill, blue eyes, bright red face and yellow legs and feet. Shape and size are identical with the Colored.

The Blue Muscovy

The Blue Muscovy no doubt resulted from a cross between the Dark and the White varieties. Like the Blue Andalusian chicken, a certain percentage of the offspring come almost black, others mixed colors and a few pure white. Although similar to the Standard Muscovys in all other requirements, its plumage should be a uniform shade of blue throughout. The exact shade of blue does not appear at present to be agreed upon and probably will remain a more or less moot question until a Standard is adopted for the variety, but a uniform shade of blue should be aimed at, this being more important than an exact shade, although a too light, washy blue should be avoided.

Weights (Colored and White) Old drake 10 lbs.; Old duck 7 lbs.; Young drake 8 lbs.; Young duck 6 lbs.

When the rain raineth and the goose winketh,
Little wots the gosling what the goose thinketh.
                                   —*John Skelton, Works.*

CHAPTER VII

Ornamental Breeds of Ducks

The Black East Indian Duck

The Black East Indian duck is one of the most beautiful of our semi-domesticated varieties. In 1943 the committee of three professional artists invited to select the most beautiful bird in the Boston Poultry Show, from a purely artistic standpoint, after giving the matter very serious consideration and after much discussion, selected a Black East Indian drake as the most beautiful bird among 5000 specimens of all varieties of land and waterfowl.

Of its origin little is known but the general supposition is that it is a development of the wild Black duck of Europe through many generations of breeding by the English poultry fanciers and is valued as an ornamental variety for ponds and streams on country estates; although it breeds well on land with only enough water to drink and in fact is more suitable for close confinement than some of the white or lighter colored varieties, as it shows little, if any, stain from earth or other discoloring matter.

The Black East Indian is prized for its small size. It is in the class with the bantam fowl as compared with the large breeds of ducks as the Rouen, Aylesbury and Pekin. The body is short, full and boat-shaped. The head is small and neat and the bill very slightly concave, not too long and well proportioned to size of head; in color it should be a deep greenish orange with purple bloom and a black bean at the end.

The neck is rather long and slender and nicely tapered from shoulders to head and with a slight curve. The legs are medium in length, set near the center of the body which should be carried in nearly a horizontal position. The color of both legs and feet should be purplish black. The color of the drake is a brilliant, green black. The higher the sheen, the better; and that of the duck shall match but with a lesser degree of green sheen. The drake is particularly brilliant on head, neck, back, wings and tail and this green sheen is, with small size, the most desirable characteristic of the breed.

While these ducks are perfectly at home on lake, pond or stream, still they will live in comfort and contentment in small yards if adequate feed and fresh water for drinking is provided. If a small swimming pool made of a barrel sawed off or a small cement pool sunk in the ground is provided, it will add to the pleasure both of the ducks and their owners as these ducks, like any other variety, in their natural environment make most interesting pictures for those who love live things in nature.

The body of the Black East Indian has a full, meaty breast and is delicious for the table. The eggs hatch well either under hen or in the incubator.

Weights: Old drake 3 lbs.; Old duck 2½ lbs.; Young drake 2½ lbs.; Young duck 2 lbs.

The Grey and White Call Ducks

The Call duck, either Grey or White, is perhaps the most satisfactory of any breed of ducks to one who wants them solely for an interesting hobby or ornament on a body of water or even a back yard or field. For it would be hard to credit this bonny little duck with much economic value. In fact, the less economic

Upper—Gray Call Drake. Lower—Gray Call Duck. (Photographs from the Bureau of Animal Industry, U. S. Department of Agriculture.)

value it has, as evidenced by size or weight, the more real value it has for the purpose for which it is usually kept and most valued. For this is really a bantam or miniature duck and like the bantam fowl its value runs in converse ratio to its size.

While not as gaudy and exotic in appearance as the more showy Mandarin or the Wood duck it is of a much more docile disposition and is hardier and more prolific. And all these qualities must be weighed against mere gaudiness when considering which variety will be most satisfactory to keep.

The Grey Call is of practically the same color scheme as the Mallard and Rouen; and the White is as the name indicates, pure white throughout.

In shape the typical Call resembles a toy duck. Its short oval boat-shaped body with short neck slightly curved, rounded cobby head, very short bill and diminutive size makes a most attractive little bird whether on land or water.

When grown, it is hardy and easy to keep. It requires no special feed and the simplest shelter is adequate. Young Calls are inclined to be delicate and should be kept out of water for the first week or two. If allowed to enter the water and stay, they will become water soaked and this brings cramps which are usually fatal. After two or three weeks however, they may be allowed to go into water freely with no bad results.

The duck will find her own nest and build it up from day to day with straws and grass until the eggs are practically hidden by the great pile of material continually being placed around it.

If allowed a good range and especially with brook or pond available, they require little or no feed from man as they are great foragers and will, during the spring and summer months, glean most of their sustenance

Upper—White Call Duck. Lower—White Call Drake. (Photographs from the Bureau of Animal Industry, U. S. Department of Agriculture.)

from the soft parts of marsh plants and grass, seeds, insects, worms, etc.

Weights are: Old drake 3 lbs.; Old duck 2½ lbs.; Young drake 2½ lbs.; Young duck 2 lbs.—or the smaller the better.

The Mallard Duck

As stated in Chapter number I, it is generally conceded by students of the origin of breeds and naturalists as well, that the wild Mallard was the ancestor of all our domestic breeds of ducks with the exception of the Muscovy positively and the Cayuga probably. The first being the domesticated Brazilian Pato and evidence pointing strongly to the Wild Black Duck of America as the progenitor of the last, possibly with some mixture of the common farm or Puddleduck.

It is common knowledge among waterfowl breeders that one may capture wild Mallards or hatch the eggs from the wild birds, bring the young up in confinement and with full feeding they will increase in size each year until in six or eight generations or even less, they will not only be so heavy they will not be able to fly but so docile in disposition that they will have little inclination, if they had the ability.

When the old birds are captured, they must however be "pinioned" by clipping the last joint of one wing to prevent their flying when the migratory urge would cause them to join the wild birds as they fly over in the spring and fall; but with one or two generations the increase in weight and the decrease in their natural migratory instinct makes the pinioning wholly unnecessary.

Varro, the Roman writer on agriculture (116-29 B.C.) as translated by Fairfax Harrison in *Roman Farm Management* says: "whoever wishes to keep a

Upper—Mallard Duck. Lower—Mallard Drake. The Mallard is a wild duck which is quite easily domesticated and which has a plumage color very similar to the Rouen. It is small in size. (Photographs from the Bureau of Animal Industry, U. S. Department of Agriculture.)

flock of ducks and establish a *nessotropheion* should
choose for it, if it is possible, a swampy location because that is most agreeable to the ducks, but if not,
then a situation sloping to a natural lake or pool, or to
an artificial pond with steps leading down to it, practicable for the ducks. The enclosure where they are
kept should have a wall fifteen feet high, such as you
saw at Seius' Villa, with only one opening into it. All
around the wall on the inside should run a broad plat-
form on which are built against the wall the duck
houses, fronting on a level concrete vestibule in which
is constructed a permanent channel in which their
food can be placed in water, for ducks are fed that
way. (For a small flock of ducks which the owner
wishes to keep in the very best of health and condition, there is no better way to feed ducks in this day
and generation.—Author.) The entire wall should be
given a smooth coating of stucco to keep out polecats
and other animals of prey, and the enclosure should be
covered with a net of large mesh to prevent eagles
from flying in and the ducks themselves from flying
out."

From this advice from Varro, one of the best of the
old Roman writers on the subject of farming, one
might assume that these ducks required the pen covered with the net to prevent their flying out and that
they were probably wild ducks; but on the other hand
it is pointed out by the same writer that the water
must be made of easy access for the ducks which would
be unnecessary were they not heavy and not too active
birds such as domestic ducks. Inasmuch as it was the
custom in those days to have all the fowls of the
farmstead confined together and that provision had to
be made to prevent the attacks of wild birds of prey, it
is logical to suppose that this net covering over all was
to protect the smaller birds such as young ducks,

chickens and the pigeons from the depredations of
eagles and hawks.

Harrison Weir in *Our Poultry* (1902) quotes from
Arthur Standish ("The Commons' Complaint," 1611)
information at some length which as he says ". . .
shows that the keeping of wild, or rather, tame wild
ducks was deemed to be both pleasing and lucrative,
and that the wild duck became tame, bred and increased to a considerable number, mixing with other
poultry; and the same way of gaining a flock of wild
or a mixture of wild and tame is in use at the present
time."

James Maine in *A Treatise on the Breeding, Rearing and Fattening of Poultry*, 1807 writes interestingly
as follows of wild ducks' eggs. "When there is a possibility of getting wild ducks' eggs, it is an easy matter
to have them hatched, by entrusting them to a tame
duck, or still better to a hen. The nests are found in
the rushes, near ponds, rivers (especially in solitary
places), in bushes that border on pieces of water fre-
quented by these birds. Nothing is afterward so easily
domesticated as the young ones coming from them;
they get accustomed to being in the midst of other
tame ducklings as soon as care has been taken to cut
the outward part of one of their wings. Without this
precaution they would fly away with wild ducks who
habitually make their abode in certain cantons, or who
pass in flocks at a fixed time of the year; but when
love has united them, they no longer think of leaving
a place that has witnessed their first affection. Wild
Ducks' eggs being to be had with the greatest of ease
in certain cantons, has induced Gouffier to propose to
economists to the renewal, every fifteen or twenty
years, of the primitive race of our ducks, by again do-
mesticating wild ducks; they are infinitely better and
cost less to keep because by nature are more inclined

than our tame ducks to seek their food; they are all
day long and at all times of year, seeking their food
by the waterside . . . The individuals of the first generation, are in truth, rather smaller than our tame
ducks but in the second, and especially the third, they
at least prove as large: they have all the delicacy of
wild ducks and all the goodness of our dabblers."

So, while there are no pedigrees and records that go
back into the mists of long ago, far enough to definitely establish the fact that the various breeds of our
domestic ducks trace directly back to the Wild Mallard, any more than there are that our domestic chickens all have descended from the Gallus Bankiva or
wild Jungle Fowl of India, the evidence we have all
points to the fact that it takes but a very few generations to develop an undersized Rouen from pure Wild
Mallards and that from these, by selection from mutations or sports, might well have been developed other
colors and types of domestic ducks which we designate
by breed names.

But that the Mallard is a hardy and really a useful
little duck, ornamental on water or land, easy to raise
and care for, and pretty well able and very willing to
make its own nest and hatch and raise its own ducklings, there is no doubt at all.

The Mallard is not a Standard duck, not having
been admitted to the *American Standard of Perfection*
by the American Poultry Association.

In color, the drake is like the Grey Call and the
Rouen and after many years of breeding for color improvement on the Rouen, it is interesting to note, the
color of the Mallard drake is just about as good as the
best Rouens. The head is the same brilliant, iridescent
green, the breast as rich in its claret and the grey body
with its fine black pencilling as delicately lovely. It is

a natural color pattern and it is hard to improve on nature's beauty.

However, the color of the duck as a rule does not come up to the perfection of pencilling we find in the best of the Rouen ducks. It is not as even nor as distinct. The Rouen excels both the Mallard and the Grey Call in the color of the female. All are rich brown, pencilled with darker brown and in varying degrees of perfection. While the Mallard resembles the Call in both color and size there is a definite difference in shape or type. While the Call is compact and cobby, the Mallard is longer on the leg and racier in outline. Its head is clean cut and racy. The bill long and more slender. The body is long and rather narrow and the whole appearance is more slender and gamey than the Call.

The Mallard is a rather shy layer although the laying tendency increases with domestication.

There is a market in some places for the Mallards for use as decoys for wild duck shooting and a market may be developed in high class markets specializing in game, but the usual use for them is for ornamental purposes on ponds and lakes.

Weights should not be over 2½ lbs. for drakes; 2 lbs. for ducks.

Goslings lead the geese to water.
—*Thomas Fuller—Gnomologia, No. 1740.*

CHAPTER VIII

Foreign Breeds of Ducks

The Duclair Duck

Here is a duck that, colored in general like the Rouen but smaller in size, and much more active, would make a grand farm duck in this country. It has never been bred in the United States. Robinson says: "According to *Vie a la Campagne* the Duclair Duck originated in the village of Duclair on the banks of the Seine, not far from Rouen, and was produced by a cross of Rouen and Cayuga. The result was a duck a little smaller than the Rouen, more active, very hardy, of as quick growth as the Aylesbury and Pekin, darker in color and not so distinctly marked. These ducks are very good layers. It may be considered a strictly farmers type of Rouen. It is said to be found nearly all over France but nowhere bred at all numerously."

Brown, however, in his *Races of Domestic Poultry* gives the following derivation for this duck, saying nothing of a Cayuga cross. ". . . we have no evidence of how it has been produced. It is parti-colored, having some resemblance to the Rouen; but as brown ducks were common at one period all over Western Europe, it is probable it has been selected therefrom. Our first knowledge of the Duclair duck was obtained from an article which appeared in the *Live Stock Journal* (1879) followed by others. From this information it was evident that the duck was exceedingly prolific and good in flesh qualities. Miss May Arnold, writing in the same journal (1880) stated that it is 'the remains of an old Norman duck, preserved by special circumstances from being crossed out by the wild duck into that duck which we have magnified into the English Rouen.' It has not been taken up for exhibition purposes and therefore is not much known outside of its own district, although it is found in other parts of France."

The Duclair is said to be of very quick growth. A good layer and free breeder. Reports are that in the United Kingdom they have proven very successful for practical purposes. They are colored like the Rouen or Mallard except they have not been brought to the same high degree of beauty as the exhibition Rouen.

Weights are: Old drake 8 lbs.; Old duck 7 lbs.; Young drake 7 lbs.; Young duck 6 lbs.

The Magpie Duck

The Magpie duck, not yet known in this country, is a comparatively new variety which is gaining in popularity in England. They are striking and attractive looking birds rather small in size, and said to be very good layers.

They are bred in either Black and White, Blue and White or Dun and White. The body is largely white and entirely so on the breast and underbody which makes them much better to pluck than the black or blue ducks as there are no colored pinfeathers on the larger part of the body.

The top of the head is black to a point just above the eyes, making a neat looking colored cap. The neck is white and the back, from the shoulders back over the top of wings to tail is black, blue or dun as the case may be which gives the bird the appearance of having a heart shaped mantle lying on the back. The pri-

maries and secondaries of the wings should be white
and the upper part of wings black, blue or dun.

For those who like to exercise their breeding ability
in producing attractive color patterns of even and uniform appearance and who wish to work with a duck
that will give economic returns at the same time, this
looks like a desirable useful and very attractive family
of ducks.

Appleyard says of this breed: "To breed a bird with
perfect markings of good sound color is not too easy;
however, this adds interest to the breed. It certainly
is a most useful duck, because it will be noticed that
the markings are so placed that when a bird is dead
and plucked there are no black or colored stubs to
show on the breast; in other words, they pluck out
just as clean as a white duck."

In shape the Magpie is rather long and racy and is
a good forager and very active. The head is long and
straight with a long nicely proportioned bill, very
slightly dished. The eyes are large and prominent;
dark in color. The neck is long and nicely curved and
the back is long and straight. The tail is carried rather
back adding to the apparent length of back. The bill
should be orange, free from dark spots, the legs should
be black and the rump, thighs and underbody, white.

Coarseness and oversize is a defect in this breed as
the Magpie is prized for alertness and a somewhat racy
appearance.

Weights are: Old drake 6½ lbs.; Old duck 5½ lbs.;
Young drake 5 lbs.; Young duck 4½ lbs.

The Merchtem Duck

The Merchtem duck gets its name from the Merchtem district of Belgium northeast of Brussels where it
has been bred by the hundreds of thousands for many

years to supply the great Brussels market. It is a purely commercial variety and resembles the Aylesbury in shape but smaller in size. It is famous for the fine quality of its flesh which it produces economically and rapidly, is very white and said to be delicious in flavor.

. The ducks are said to be most prolific layers of nearly white eggs. Many of these ducks in the Merchtem district are reared in enclosed yards with no access to water except for drinking purposes. The birds fatten rapidly and can be marketed as green ducks at a very early age.

Sir Edward Brown in *Races of Domestic Poultry* says of its origin: "The descent of this breed of ducks appears to be very uncertain. The general opinion is that it is a sport from the Blue Termonde but smaller in size. In appearance it greatly resembles the Aylesbury except it is not so large, and the bill often has a blue bean; also the legs are blue. These differences give it a distinct character and all suggestions that the Aylesbury and the Merchtem are one and the same may be dismissed as incorrect, although there are floating traditions that white birds were formerly exported to England."

In body the Merchtem is long and boat shaped with the lower part of the breast, or keel, being close to and parallel with the ground. The neck is medium in length and well arched, the head fine and the bill long and straight and pinkish white in color. The legs set nearly the center of the body, are orange in color.

The plumage is white with a tendency to a bluish tint and perhaps bluish-white would better describe it. There is not the slightest tendency to creaminess as in the Pekin.

Weights are: Old drake 6½ lbs.; Old duck 5½ lbs.; Young drake 5½ lbs.; Young duck 4½ lbs.

The Termonde Duck

This is another Belgian breed and a useful and attractive one. It is one of the largest of all the European breeds ranging with the Aylesbury and Rouen and is a big, sturdy duck not as fast growing as the Pekin but a fine layer and very vigorous.

Referring to the strength and virility of the Blue Termonde, Sir Edward Brown in *Races of Domestic Poultry* says: "The suggestion is made that the reason why the Blue Termonde duck has attained so large a measure of popularity is that birds with this color of plumage were less seen by their enemies, both human and animal, than if they were black or red and so were preferred by the peasants. But there is another cause, namely, blue plumage in ducks represents influences which make for vigor of constitution and quality of flesh."

Whether this last is so or not is a matter for discussion but that the neutral blue-grey color would seem to make better camouflage for the low lying marshes and swamps of Belgium where these blue ducks were raised, sounds reasonable.

The Blue Termonde is very long in body and deep; with breast carried well forward and boat shaped. The neck is long and nicely arched and the head long and carried well back. The bill is long and blue in color. The eye is dark and rather full and not set too high in the head. The legs are longer than in the Aylesbury and set well back, giving the bird a rather upright carriage. The legs are slate colored.

The plumage is a blue or slatish grey, each feather slightly bordered with a darker shade. Sometimes the breast is broken with white feathers and even in some specimens a white bib appears, as in the Blue Swedish, to which in my opinion, there is more or less direct relationship.

Brown says, of the flesh qualities of the Termonde: "The great value of the Blue Termonde is in crossing with softer fleshed ducks, as in the case of the Indian Game and Dorking Fowls, and it secures this result without the use of yellow fleshed ducks so, that the color of the skin and flesh is not adversely affected while the quality of the flesh is greatly improved. It is a bird of great vigor, hardy in the extreme and the ducks are excellent layers."

Weights for Old drake 10 lbs.; Old duck 8 lbs.; Young drake 8 lbs.; Young duck 7 lbs.

What meaneth he by blinking like a goose in the rain?
—*William Bullein, A Dialogue Against the Fever Pestilence, 1564.*

CHAPTER IX

Breeding Ducks

Whether the best results come from breeding from yearling or from older ducks will probably never be decided. One breeder is sure, from his own experience, that the best and strongest ducklings come from yearling breeders in the pink of high condition and the next duck grower, from an equally long and broad experience, is just as sure that the very best ducklings come from breeders two years old.

This question has never been decided in the field of hens and probably never will and so there is no hard and fast rule, either with ducks or hens. If the young birds are well grown and well matured when the breeding season comes nothing is gained by waiting. It looks like the Almighty planned the breeding season and the mating instinct to synchronize in ducks as well as chickens, if hatched during the natural breeding season and so if the ducks are hatched in late winter or early spring they should be at the very acme of breeding condition about the same time the following year.

However, there is a natural mortality during the first season and naturally whatever birds do die are apt of course to be the less hardy specimens so that the average health and strength of the flock at the end of the second year may be expected to be a little better. Thus the inherited tendency of the young from older parents to live, may be to that extent higher.

But if good strong, vigorous ducks are selected and mated with large active and virile drakes, there will be found little if any difference whether the breeders are

A good house for breeding ducks. It is 20 feet deep, 40 feet long, 7 feet high in front and 4 feet in the rear and will accommodate 200 breeders. (Photograph from the Bureau of Animal Industry, U. S. Department of Agriculture.)

one or two years old. If wintered over the second year care should be taken however, that they are not allowed to become over-fat as too-fat breeders will cut down the fertility and hatchability and so materially reduce the profits.

Whether for meat, or eggs only, the most vigorous and active ducks should be selected for breeding. Select for health and vigor first and then for breed-type, temperament, size and color. It will be found that whatever the purpose, the specimens within the breed that come closest to the type and size that has been found by long experience to best fit the purpose for which that breed is intended will bring, as a rule, the best results. Select for the following qualities in the following order. Vigor, temperament, correct size, true breed-type, approved color.

With these in mind, and in their relative importance, you will make no mistake. Bear in mind always the law of compensation. There are two parents to every young one, and if the drake is lacking slightly in one characteristic of size, or breed-type or correct color, select for the female side specimens strong in the points in which the drake fails. But, for health and vigor there can be no "compensation." Both sire and dam must be strong in these all essential characters or the task of producing really fine specimens is hopeless.

I have mentioned temperament and perhaps my reader is wondering why. Let me explain the quality of temperament. By it is meant the nervous quality that makes the specimen better fitted to perform the duty for which it is intended. Thus, a meat animal should not be high strung and nervous. It should be of a somewhat phlegmatic disposition, not easily frightened or disturbed and capable of taking things easy and assimilating its food efficiently without "burning up" too much food-fuel in unnecessary activity.

Interior of house for breeding ducks. Notice the heavy bedding and the feeding track. (Photograph from the Bureau of Animal Industry, U. S. Department of Agriculture.)

On the other hand, the egg-bred duck should be of a higher tension and more nervous temperament. It has been proven both in hens and ducks that high egg production and a fairly high-strung disposition go hand in hand. Note the difference between the nervous, active Indian Runner or Khaki Campbell and the stolid dispositioned and phlegmatic Rouen and Aylesbury. So, after the quality of vigor I should place "temperament" in the order of importance when selecting breeding ducks.

Having in mind the factors necessary to reproduce in the young to make your breeding work successful, it remains to select the individuals combining those factors in the highest degree. For no matter how large the farm nor how many ducks there will be in your breeding flocks, every mating consists of just two individuals and thus two individuals multiplied many times or few, depending on whether your breeding work is on a large or small scale, will absolutely determine the average inherited quality of the young flock you raise. And on the quality of the young you raise will very largely depend the profit or loss in your duck enterprise: so you can see how very important it is to select with the utmost care, the breeding units with which you will work.

First, let us consider the drake. He is as important as five or six females, as he will mate with five or six; and so take plenty of time and care in selecting him. If selecting drakes from a flock, first eliminate those that lack in the fundamental qualities of vigor, size and temperament; study the flock and take out all those that plainly lack all or any of these important characteristics. Then, proceed to weed out those lacking in proper breed-type or correct color. Then from these which are left select the one individual which best combines all these qualities and after him select

Two methods of carrying ducks. (Photographs from the Bureau of Animal Industry, U. S. Department of Agriculture.)

the one closest to him and so on till you have the required number. Do not depend on your eye. Handle each bird to determine whether he is well fleshed and solid. Whether his breast is broad and deep, heart-girth round and full, his body wide and long and whether his eye is full and bright, his legs strong and set wide apart with plenty of room for breast and body; and his whole "feel" strong, live and springy.

Size is important but not excessive size. A little over Standard weight is desirable as the tendency is toward mediocrity in size or a little under and so it is well to breed from parent stock a little over the Standard weight—but not too big. Drakes too much oversize will be apt to be less active and fertility from them not run as high.

Follow the same rules in selecting ducks. Bear in mind whether for meat or eggs we must have vigor, health and the correct temperament for the purpose intended. The meat ducks should be rather quiet and docile in disposition and the egg breeds more active and while not "scarey" or too high-strung, they should appear alert and attentive to what is going on around them.

Legs and feet are important. The legs of the wild duck were never intended for much travel on land. Flight and swimming were nature's mode of transportation. In domesticity the body has doubled in size and there is more or less walking on land. Far more than nature intended. So, it is most important that special attention be given to legs and feet when selecting ducks for breeding. A breeder is no better than its legs and feet and no matter how fine the body and how many eggs the duck has within her if she "goes down on her legs" she is a loss. Walk every breeder forty or fifty feet and see if they travel with ease and confidence, when selecting. Heavy ducks waddle of course,

Upper—Comparison of size of goose egg on the left a black egg of a Cayuga duck in the center and a hen egg on the right. Lower—Duck eggs—At the left is a Pekin duck egg, next a black egg laid by a Cayuga duck, third a Muscovy egg, fourth a duck egg of green color and on the extreme right the egg of a Runner duck. (Photographs from the Bureau of Animal Industry, U. S. Department of Agriculture.)

but it's easy to see whether this is caused by weight
on sturdy, wide spread legs or whether by legs that
are weak or improperly set. Legs and feet of course
are equally important on the drake.

Plenty of time spent in studying the candidates for
the breeding pens of ducks will pay big dividends and
should never be slighted.

It is better to make the first selection from which
the final choice of breeders will be made before starting
to fatten the young ducks for market. Fattening ra-
tions are largely composed of corn which is heating
and produces much internal fat at the expense of the
stronger frame development, as when the birds are
grown along the natural way.

Forcing for heavy early market condition either in
chickens or ducks is apt to be detrimental by preventing even natural growth of strong frames and the normal development of the reproductive organs to the
condition necessary to safely stand the tremendous
grind of heavy and sustained egg production through
the breeding season.

Make the first selection of breeders at about the
eighth week. Then inspect again at about ten weeks
old or the normal time for marketing, had they been
developed on fattening ration. Again at four or five
months the young birds should be gone over and a
semi-final selection made. At this time the young
drakes should be separated and kept apart till the final
selection for the breeding pens is made, at about six
to seven months old.

Breeders should be not less than seven months old
when they come into egg production. Earlier laying
tends to limit size and while perhaps it does not hurt
health and production, it certainly does not improve
either one.

Figure back approximately seven months from the

Meal time for the breeders. (Photograph from the Bureau of Animal Industry, U. S. Department of Agriculture.)

time you expect the young breeders to lay and the hatches from which you desire to select breeders can be given special attention while developing. The young birds to be used as breeders should be from the best hatches from the best flocks. Eggs from which you intend to select future breeders should be selected with utmost care for uniformity of shape, size and shell texture for all of these things are hereditary and "like begets like," with natural variations.

Mating

Under average conditions and with the common breeds from five to six ducks in the breeding flock is about right. This varies a little with the climate; the colder climate and earlier in the season requiring more drakes than in the more moderate climate and as the season advances. It is safer to use a higher percentage of drakes in the northern climates especially in the early season even if some culling is done as the weather warms up and less drakes are required.

If breeding is done late in the season as the weather grows warm, more drakes may be again needed, so in mid-season when drakes are removed to prevent injury to the ducks by excessive treading, they should be segregated for use again later on if fertility begins to wane with the hot weather.

The age of the drakes should equal that of the ducks and even a month older may be advisable. It can do no harm. When the drakes are selected for breeding and segregated from the ducks at about four or five months, numbers should be carefully checked that there may be enough for use when the pens are mated, and it is good practice to keep a pen of extra drakes to use in case of accident or sickness of any kind.

Incubator cellar on large duck plant. Trays of eggs set out to turn and cool. (Photograph from the Bureau of Animal Industry, U. S. Department of Agriculture.)

CHAPTER X

Hatching

The period of incubation for duck eggs is 28 days with the exception of the Muscovy or Brazilian which is from 33 to 35 days. Like hen eggs, the eggs of the duck may be hatched either naturally or by artificial methods and like hens, the breeds that have been developed to a high degree of laying are practically non-sitters.

If hatching but a few ducklings, the hen or duck method will probably be most satisfactory for by this method the brooding problem is also solved. The hen as a rule will own them and will care for them just as well as if they were chickens. A hen will cover from 9 to 11 eggs and the duck from 10 to 13. The Muscovy is a natural mother and will hatch and brood her own ducklings—or those from other breeds of ducks—with little or no assistance. Let her choose her own nest and when she becomes broody just supply the eggs (if she has not stolen the nest and laid a nestful before you find it, which she is very apt to have done) and she will do the rest. All she needs is some feed once a day and water to drink when she leaves the eggs for the daily airing.

In artificial incubation the same principles apply to both ducks and hens eggs except that the temperature should be about 1° lower for ducks than for hens and the humidity should be a little higher. It has been found that the small laying ducks such as the Khaki Campbell and the Runners require about ¼° higher temperature than the large breeds.

In still air machines the correct temperature is about 102 and in forced draft incubators about 100°. It is impossible to give specific directions for the hatching of either ducks or hens eggs in a work of this kind. There are so many different kinds of incubators and they are run under so many different conditions of housing, temperature and humidity that the safe and sensible thing to do is to follow as closely as possible the directions of the manufacturers; and then these may have to be adapted to your own conditions. Temperature readings in the still-air machine should be made with the bulb of the thermometer 2 inches from the floor of the tray. The temperature of the hatching compartment should be varied slightly according to the temperature of the incubator room. A proper temperature for the incubator room is 60° F. If the room is kept as low as 40° F., the thermometer in the machine should be $\frac{1}{2}$° higher and if the room temperature is as high as 80°, run the incubator temperature $\frac{1}{2}$° lower. If an agitated air machine is used, however, do not vary the incubator temperature more than $\frac{1}{4}$ degree.

Care of Hatching Eggs

Plenty of nests with clean, dry straw for bedding, reduce the number of cracked and soiled eggs. Ducks usually lay in the early morning and the eggs should be gathered as soon after laying as possible. Wooden or galvanized iron pails with cloth or burlap in the bottom make good receptacles for gathering the eggs. Care should be taken in very cold weather to prevent the eggs from chilling while being carried from the laying houses to the incubator cellar. A piece of blanket over the pail to protect it from the cold wind will help. During extreme cold weather, line the pails with heavy paper or cloth.

Clean badly soiled eggs as soon as possible. The sooner the eggs are cleaned the easier it will be to clean them and filthy eggs should never be set. The water should be somewhere near the same temperature as the eggs when washing.

The hatching quality of eggs decreases steadily with the time they are kept. If necessary to keep more than a very few days, eggs should be stored in a room with high humidity and an even temperature; as near 50° F. as possible.

In large duck breeding enterprises a special room for storing hatching eggs will prove a good investment. Under no circumstances keep eggs in a warm place but if proper temperature and humidity can be maintained, they may be stored for 10 to 14 days without serious deterioration in hatchability although a safe rule is to set them just as soon as possible after they are laid.

From the fact that duck eggs require more moisture than hen eggs, it is very difficult to hatch both in the same machine at the same time. Humidity during the first 24 days should be from 55 to 60 degrees and from then to the end of the hatch from 55 to 60 degrees for duck eggs. This applies to all types of incubators.

In forced draft machines with separate hatchers, the wet-bulb reading for the first 24 days should be 89° to 92° F. and during the last four days from 85° to 87° F. The wet bulb measure for humidity cannot be successfully used in the still air type of incubators. In small incubators, after the first week, moisture is usually provided by sprinkling the eggs daily with luke warm water or by placing a pan filled with warm water or wet sand below the egg tray. Soaking the floor of the incubator room is also sometimes helpful. No hard and fast system can be laid down here as conditions of humidity vary so in different incubator rooms.

Eggs in the flat tray incubators should be turned at
least twice each day and as many as three or four times
if possible. In agitated air machines with automatic
turning devices it will be advisable to turn the eggs
three or four times each twenty-four hours. There
should be but a moderate supply of fresh air up to the
twenty-fourth day but for the balance of the hatching
time plenty of fresh air is very desirable. Eggs should
be tested on the seventh and the twenty-first days.

Brooding Ducklings

As in the matter of incubation, either the natural or
artificial method of brooding may be employed. The
natural method is to brood with either hens or ducks.
Many ducks, especially in the heavy laying strains, do
not sit and so manifestly are not available for brooding.

But, if you have ducks that will sit and you are
raising ducks in a small way, the "duck-incubator-
brooder" combination is the most satisfactory as a
rule. Either a good sitting hen or duck will hatch a
high percentage of fertile eggs and then will raise the
ducklings with very little trouble. If possible, let the
duck find her own nest and when she shows she really
means business by sticking to the nest and ruffling her
feathers and hissing when approached, slip 10 to 13
eggs (depending on comparative size of eggs and duck)
under her and let nature take its course. See that she
has some whole corn if possible once a day and that
there is water available for her to drink and if convenient, for her to bathe in once a day, and there is little
to worry about until the ducklings hatch. When
hatched, it is best not to let them have water to swim
in for the first few days as they are apt to get their
down soaked, with bad results. Cramps are caused by

Interior of brooder house showing walk and hover combined in the middle of the house and pens on each side. (Photograph from the Bureau of Animal Industry, U. S. Department of Agriculture.)

soaking and frequently result in death. After the second week, there is little danger from this and they may have free access to the water, if available.

If using hens for brooders, it may be a little difficult if the hen does not hatch the brood of ducklings. After all, there is considerable difference in appearance between the ducklings and chickens and unless the mother hen gets acquainted as they come from the shell, she may not own them. If the baby ducks are given to a hen to brood, place them under her in her coop at night and as a usual thing she will own them when light comes in the morning. If you have no sitting hen or duck available and you buy baby ducklings, use a small lamp brooder or one heated with an electric light bulb or even a box with a bulb suspended from the closed top will usually be sufficient to heat the little ducks for the first week or two as they require heat only a short time except in the very cold weather and if only a few are raised it is well to wait till later in the season before getting them.

After the first week they are much sturdier than chickens and do not require as much heat nor do they require it as long.

When hatched in incubators and brooded artificially, they should be placed in the brooder for 24 to 36 hours after hatching and not fed until then. On larger scale operations, from 100 to 150 ducklings are usually placed in a pen or colony house of about seventy or eighty square feet with a hover-type brooder heated by either oil, coal or electricity. The temperature under the brooder should be about 95° the first week, 85° to 90° the second week, 70° to 75° the third week and 70° thereafter, depending of course on the climate and the outside temperature.

Observation will tell you how long the artificial heat must be available. When the ducklings cease to go

Another type of brooder house. Here the hovers are along the back of the house and the work is done from an alleyway along the front. The box with handles on top of the hover is used in carrying the newly hatched ducklings from the incubator cellar to the brooder house. (Photograph from the Bureau of Animal Industry, U. S. Department of Agriculture.)

under the hover, when resting, there is no more need of artificial heating.

When first placed in the brooder house, they should be confined near the hover by a circular fence of 12-inch wire netting, cardboard or other material until they learn to go under the brooder when requiring heat. Then, it may be removed and their own comfort will tell them when to seek the heat; and they will have learned where.

Under-hover temperature is all important. The temperature in the brooder house may vary to a considerable degree if there is warmth enough under the hover so the ducklings may warm themselves and be comfortable at any time. But heat for resting and in the proper degree is essential to satisfactory growth.

On the big commercial duck farms, the ducklings usually are taken directly from the incubators to what is known as the "hot house" with room temperatures of 70° the first two weeks and then transferred to one with lower temperature and from that after a couple of weeks to one with no heat.

But before the hatching season starts, yes, the end of the previous year's brooding is none too soon, thoroughly clean, renovate and disinfect the brooder house for complete readiness for its next use. Floors should be scraped and scrubbed, all manure and litter carted away. The floor well soaked with disinfectant and walls and ceilings sprayed with the same.

Hovers, feed hoppers and waterers should be scraped and scrubbed and put out in the direct sunlight, the great sanitizer. Open up the windows and let the interior have the same direct sunlight and fresh air.

After the house has been thoroughly cleansed and disinfected, cover the floor with five inches of clean, dry sand. Starting the ducklings in a clean sanitary

Brooder house and yards. The trees furnish fine shade for the growing ducklings. (Photograph from the Bureau of Animal Industry, U. S. Department of Agriculture.)

house will help much to give them the good start that is all important. Right through the brooding season, as each hatch is moved out, the brooder houses should be cleaned and disinfected before the new brood is brought in.

CHAPTER XI

Feeding

Since ducks are greedy feeders and as the feed is by far the largest single expense in raising or keeping them, it is most important that suitable feed be provided and that it be provided in such manner as to have as little waste as possible. For ducks are not only greedy but slovenly and wasteful of their feed if precautions are not taken to prevent.

The wild duck never has access to masses of food. It hunts its ration on low marshy lands and in the water of streams and ponds, taking a little here and a bit there and so there is little if any waste; but in domesticity feed must be furnished and as ducks thrive much better on ground feed than on grain, it is obvious that it must be fed in some kind of receptacle from which the duck will throw out and waste as little as possible.

Unlike the goose, the duck is an omnivorous feeder. Grain and green food and animal matter are all needed, and in the right proportion; in this respect the duck is like the hen. In the wild state both will hunt their food and will find it in small amounts here and there, and seeds, grains, green food, worms and insects all go to make the balanced ration which is so necessary for them to thrive and breed and reproduce: and in domesticity we must provide this balance.

When the young ducks are hatched, they require nothing for the first twenty-four to thirty-six hours. They should be kept warm and comfortable until the yolk from the egg which is taken into the body the last thing before hatching, has become thoroughly assimi-

Brooder house. At the time this picture was taken there were no ducklings in the house and advantage was taken of this fact to give it a good cleaning by throwing out the bedding and droppings, which will be hauled away and spread on cropped land. (Photograph from the Bureau of Animal Industry, U. S. Department of Agriculture.)

lated. The young ducklings should be kept in a warm place—if in brooders the temperature should be at about 95° under the hover and 70° outside in the house—and will grow constantly stronger from the nourishment derived from the absorbed yolk, up to 24 to 36 hours.

The first feed may consist of a mash composed of a mixture of two parts of wheat bran and one part of corn meal to each quart of which should be added a small handful of sharp sand or fine gravel and the whole moistened to a rather soft consistency with skim milk, if milk is available. If not use water and put in a small handful of fine, high quality beef scrap to each quart.

Feed this four times a day: oftener if convenient. Give the first feed early in the morning and the last just before dark with the others spaced at equally distant times between. This may be left before the ducklings but I like to feed them just what they will eat quickly at one time and then take the receptacle away until time for the next feeding time arrives.

For the first two days under the hover it is well, however, to leave the feed by them in small troughs or feeders, constantly. By this they will learn to eat more readily and at the end of two days their appetites will be developed to the point that they will stuff themselves and then between drinks will digest the food quickly and be ready to again eat their fill when the next feeding time comes around

It is most important that water be kept by them constantly. A duck requires an abundance of water for drinking; they almost need a drink of water after every mouthful of food. The water must be fresh and cool and nothing will do more to keep your young ducks growing and in good health than plenty of fresh water constantly before them. Place the water dishes some distance from the feed. The exercise will help the youngsters' growth.

The same mash may be fed until after the sixth or seventh day, gradually increasing the beef scrap in the mash until at five weeks old they are getting from 10 to 12%. A little low grade flour may be added to make the mash hang together and prevent waste as far as possible by the ducklings throwing it about while eating.

Feed them but three times a day after the fifth week and only what they will eat up clean at each feeding. After the first three weeks, green food such as clover, fresh green grass, rye or oats should be cut short by running through a hay cutter and mixed with the wet mash each day.

Fine gravel or grit should be before them at all times and much care should be taken that they are not startled or frightened. Young ducks in flocks are exceedingly timid and nervous and sudden starts do the growing birds harm and retard their growth. Quiet is a most important part of the schedule for growing ducks economically and well. It is not necessary that the green feed be added to the mash unless the birds are being raised for breeders although if grown for meat it will cut down the feed cost very materially.

In the early spring, the growing ducklings may be removed to cold houses at five or six weeks of age, depending on the weather, as they will grow better, as their feathers grow and the weather gets warmer, without artificial heat.

Later in the season the ducklings may be removed at three weeks of age. This of course when summer weather comes.

There are many other systems of feeding and many different ingredients used in varying proportions. Practically every large duck grower has his own special

Watering arrangement in the brooder pens for young ducklings. (Photograph from the Bureau of Animal Industry, U. S. Department of Agriculture.)

little hobby or personal system of feeding but the above has been proven by long experience to do well if adhered to. But that is just in case you cannot get the greatest boon yet provided for the duck grower by the better commercial feed mills. I refer to pellets. These are scientifically compounded for the different requirements of duck nutrition; starting, growing, fattening, laying, breeding. While pellets cost slightly more per pound than the mashes, they more than save the extra cost from the fact that the ducks waste so much less than in the feeding of mash. A duck will take a pellet and swallow it, take a sip of water and then another pellet and so on. With a mouthful of mash there is always some waste as the duck goes for the water to wash it down.

The best advice I know is to feed pellets, and only pellets, unless it is handy to feed some green food in addition to the pellets. With pellets, the best system is to keep them constantly by the growing ducks and with the fresh water always at hand. But not too near at hand. A good plan is to keep the hoppers with pellets in the house and the water fountains out of doors just as soon as the ducklings are old enough to be out all day. This will give the birds exercise, which is very important, as they will run from the feed to the water and thus not only keep the litter inside of the house dry but the ducklings will have the benefit of sunlight and fresh air. Of course while the ducklings are still young, an indoor fountain will be required and kept constantly filled.

After the ducks are old enough to remain out doors day and night, large feed hoppers with weather-proof roofs will save much labor by only requiring replenishing at intervals, as a good sized one will provide food for the flock a number of days.

Ducks being raised for market should be gradually

Upper—Pekin ducklings 3 days old. Lower—Pekin ducklings 2 weeks old. Duck egg used for size comparison. (Photographs from the Bureau of Animal Industry, U. S. Department of Agriculture.)

Upper—Pekin ducklings 3 weeks old. Lower—Pekin ducklings 6 weeks old. (Photographs from the Bureau of Animal Industry, U. S. Department of Agriculture.)

changed from the growing ration to the fattening ration at about seven to nine weeks of age, depending on size, and at what weight you plan to market them.

Weather conditions affect the growth of young ducks. Cold, rainy weather or that which is unseasonably hot, tends to retard the growth regardless of feed. Lack of shade in the hot summer weather will cause serious trouble and loss in growing ducklings. Overheating causes leg-weakness, giddiness, whirling spells and spasms often resulting in death or at least badly stunting the young bird. Shade is important, from which they may seek shelter from the sun, but they must also have as much sun as they desire.

In feeding young birds for breeders the feed should be changed at about thirteen weeks old or the time they moult and to the time they begin to lay, feed them some grain with the pellets or mash. Body strength and hard flesh is important rather than the soft tender fat and flesh desired for market. They should have exercise and be made to forage by cutting down feed if necessary.

CHAPTER XII

Marketing

When young ducks have reached the age of approximately 12 weeks, the flight feathers have just about reached full length and they are well fattened, they have reached the period known in marketing parlance as "green ducks." This is the stage at which they bring the best price on the market and the time when most market ducks are sold.

Green ducks are sold dressed as a rule, although the demand for live ducks, especially on the New York Market has grown greatly within the last few years. Pekin ducklings at the "green duck" age should weigh about 5½ lbs. as this is the weight most in demand and which brings the most favorable price.

While for the nearby markets there seems to be an increasing demand for live ducks, live ducks do not stand shipping well and the custom will probably never increase to a great extent except to nearby points where the shipping time is very short. In locating a duck farm proximity to market is very important as in any other kind of agricultural or live stock production.

Ducks are usually hung by the feet for killing and the jugular vein cut from the front of the mouth with a sharp knife. This vein is located at the base of the skull at the rear of the throat and a little practice will enable one to hit it with little trouble. Ducks should not be fed for at least twelve hours before killing as food in their crops renders them unsightly and hurts the keeping qualities.

There are two methods of plucking. Dry picking

Carrying the ducklings from the catching pen to the killing place. (Photograph from the Bureau of Animal Industry, U. S. Department of Agriculture.)

and scalding; the most of them are scalded except in some markets where the demand is for the dry picked birds. The water for scalding should be just below the boiling point or about 190° F. If too hot it will discolor the flesh and render it unsightly, thus hurting the price.

Ducks should be scalded and plucked as soon as possible after bleeding. If only a few ducks are to be scalded at one time an ordinary wash boiler may be used but if many are to be plucked a vat of some kind should be prepared and thermostatic or other automatic means of controlling the temperature of the water be provided. Scalding vats are manufactured and sold by those manufacturing general poultry equipment.

The method used is to take the bird by the head and feet and souse it into the water, holding it there long enough to let the water penetrate into the breast and body feathers to the base. When the feathers may be removed easily the bird is sufficiently scalded. If the head and legs are held out of the water the bill and legs retain the bright yellow color and are much more attractive when marketed.

In picking, start with the breast and body and work toward the tail, holding the body of the bird in the lap or on a table. The long tail feathers are left in as a rule, the wings are picked to the first joint and the neck half way to the head.

The most difficult part is removing the down. This is accomplished by rubbing, but care must be used not to break the skin. Sometimes the down is shaved with a very sharp knife. Sometimes the duck is sprinkled with powdered resin and then dipped into hot water, thus melting the resin, when down and resin may be rubbed off clean. The long pinfeathers are removed by

The ducks are hung by the feet and the veins in the neck cut from inside the mouth to cause free bleed-ing. (Photograph from the bureau of Animal Industry, U. S. Department of Agriculture.)

catching them between the thumb and the blade of a knife.

Picking ducks quickly and well, requires much experience. Good pickers on the large duck farms average as many as seventy-five ducks a day. Immediately after dry picking, they should be washed and chilled in cold water until the animal heat is removed. They are then again chilled in ice water or a cooler and at once packed for market. The ducks are usually packed on their sides or breast down, between layers of ice in barrels. On the top of each barrel is left enough room for a layer of cracked ice on top of the packed ducks. Boxes are also used for packing and it is well to grade the ducks according to size that they may present a uniform appearance when sold.

Wax Method of Plucking

The wax method of picking has been used for some time and quite extensively for chickens but not to any great degree with ducks.

If this method is to be used the ducks should be scalded at a temperature of 190° F. to loosen the feathers. Then, after rough-picking the body feathers and removing the wing and tail feathers completely, the body feathers and down remaining should be thoroughly dried with a fan before applying the wax.

The ducks are then dipped in the melted wax twice and then plunged into ice cold water to chill and harden the wax. Following this, the wax coating should be removed by first breaking the covering on legs and breast and then rolling wax, feathers and all, from the rest of the carcass. In this way feathers, down and pins are taken off with one operation and the skin left smooth and clean.

Special wax for this purpose is manufactured com-

After the throat veins are cut, the ducks are allowed to hang until they are well bled out. The blood is caught in the trough below. (Photograph from the Bureau of Animal Industry, U. S. Department of Agriculture.)

mercially and may be bought from dealers in poultry supplies and equipment. The greater part of the wax may be remelted, recleaned and reused.

A thermostatically controlled electric unit is most satisfactorily used in melting the wax and maintaining the correct degree of heat. A very high scalding temperature should be used in order to cut the oil in the feathers, that the wax may stick.

Marketing Duck Eggs

Duck eggs for market should be stored in a cool, damp place with a temperature from 40° to 50° F.

For profitable egg production ducks should be hatched at the proper time and properly fed and managed during the growing period as well as while laying. It is just as important to get high production during the period of high egg prices with ducks as it is with hens.

It is impossible to get satisfactory winter egg production from late hatched stock. Usually the best results will come from March, April and May hatched ducks and it will probably pay best to market the laying ducks in late spring or early summer before they begin to moult. Excepting of course, those that may be reserved for breeders the following year.

The market for duck eggs is not as general as for hen eggs but is steadily expanding. In New York, the price is highest during the fall and winter, gradually reaching the top at Easter time as a rule. Usually they do not go much over the price of the best hen eggs except at Easter time and then they exceed it by from ten to twenty-five cents. (See table on page 326.)

Mr. A. C. Dingwall of the *Egg and Poultry Review* writes as follows with reference to the table which you see on page 326: ". . . there is no hen egg quo-

Ducks which have been bled, ready to have the blood washed from their heads and mouths before they are picked. (Photograph from the Bureau of Animal Industry, U. S. Department of Agriculture.)

tation that is comparable with quotations on duck eggs. Through a single year, or over a period of years, it is impossible to select any one grade of hen eggs that will correspond with the average grade of duck eggs. The best that I have been able to do is to pick out one standard grade—fresh extras, large, that weigh 45 pounds or more to the case. This grade of hen eggs, I would say, is better than the average grade of duck eggs. However, you will be able to make a comparison that will give you an index of the relationship between the prices of duck eggs and hen eggs."

Because duck eggs are larger than hen eggs, special cases and fillers are required. But, by leaving the diagonal section of each filler empty, ordinary cases and fillers may be used, carrying twenty-five dozen duck eggs instead of thirty dozen hen eggs as is customary. However, it is necessary to nail a half inch strip across the end on top and across the center before nailing on the cover.

Selling Duck Feathers

From five to seven ducks will yield a pound of feathers. At the present time duck feathers are extremely high due to the great number used up in the war and are bringing about 90¢ a pound. In more normal times the price is from that down to as low as thirty-five cents. It may readily be seen by this that the feathers form no small part of the profit and in many cases pay for the cost of picking.

On the big duck farms of Long Island, it is customary to gather all the feathers picked each day to be delivered at once to the central processing factory. Here they are cleansed, dried and graded for the feather trade. The prices quoted are for the down or underbody feathers as the soft feathers are most in

After they are bled and washed, the ducks are laid in the picking room ready for the pickers. (Photograph from the Bureau of Animal Industry, U. S. Department of Agriculture.)

Holding the head in one hand and the feet in the other, the picker dips the duck in water heated nearly to the boiling point and souses well to work the water into the feathers until they pluck easily. (Photograph from the Bureau of Animal Industry, U. S. Department of Agriculture.)

demand and bring the best prices. The quills have very little value.

Feathers from dry picked ducks are most valuable as certain deterioration takes place from scalding and wetting. However, practically all the large duck farmers in the great commercial duck growing sections of Long Island scald their ducks.

On farms distant from the processing factories, where feathers are saved for sale, the coarser feathers should be separated from the finer and spread two or three inches deep on the floor of a loft or light, dry building where it is airy and where they may be stirred up and re-spread every day. This should be continued until they are perfectly dry when they are ready to sack and ship to the buyers. They should be thoroughly dry, as if there is dampness they will heat and rot in the sacks.

Duck Manure

Another by-product from which there should be some income is the manure. Duck manure is a source of organic matter, valuable for any object for which animal manure is used. It is about twice as rich in nitrogen as ordinary barnyard manure and contains six times as much phosphorus and has fully as much potash. The average analysis shows 22 lbs. of nitrogen, 29 lbs. of phosphoric acid and 10 lbs. of potash to the ton.

COMPARATIVE PRICE OF FRESH HEN EGGS AND DUCK EGGS ON THE NEW YORK MARKET

(Hens Extra Large — 45 lbs. or more)

URNER-BARRY QUOTATIONS

	Hen 1941	Duck 1941	Hen 1942	Duck 1942	Hen 1943	Duck 1943	Hen 1944	Duck 1944	Hen 1945	Duck 1945	Hen 1946	Duck 1946
January	20.9	26.5	35.4	38.67	41.2	42.90	37.2	51.8*	45.1	56.0###	41.2	48.7#
February	18.5	30.8	30.9	41.28	38.0	43.00	35.0	47.3*	40.6	56.8####	36.0	44.8#
March	20.4	41.7	29.8	47.45	40.0	46.83	33.9	49.2*	38.8	62.5####	37.2	50.2##
April	23.6	36.1	31.6	40.14	38.9	56.68	33.4	51.5***	38.8	55.4####		
May	25.0	26.48	31.9	32.95	39.8	48.95	33.0	31.2**	38.8	50.4#		
June	27.5	28.19	32.6	34.05	41.2	42.77	35.2	29.1**	40.6	57.7#		
July	28.2	29.05	34.7	35.70	45.5	44.78	41.0	31.3**	43.6	53.1#		
August	30.9	29.57	37.1	39.10	48.6	48.77	43.4	31.4*	47.2	51.0#		
September	33.3	32.14	38.9	42.43	52.6	53.20	48.8	39.8*	44.0	50.1#		
October	36.4	35.27	42.9	43.00	55.5	60.32	51.3	45.9**	50.9	52.1#		
November	40.1	40.06	43.0	43.00	52.9	65.04	52.8	55.1**	52.8	55.4#		
December	36.7	38.00	43.0	43.00	45.4	58.1*	51.7	56.0**	51.7	57.9#		

* 1 to 3¢ premium for fancy heavy quality
** 1 to 5¢ "   "   "   "   "
*** 1 to 6¢ "   "   "   "   "
\# 2 to 6¢ "   "   "   "   "
\## 2 to 7¢ "   "   "   "   "
\### 2 to 8¢ "   "   "   "   "

Picking the ducks. (Photograph from the Bureau of Animal Industry, U. S. Department of Agriculture.)

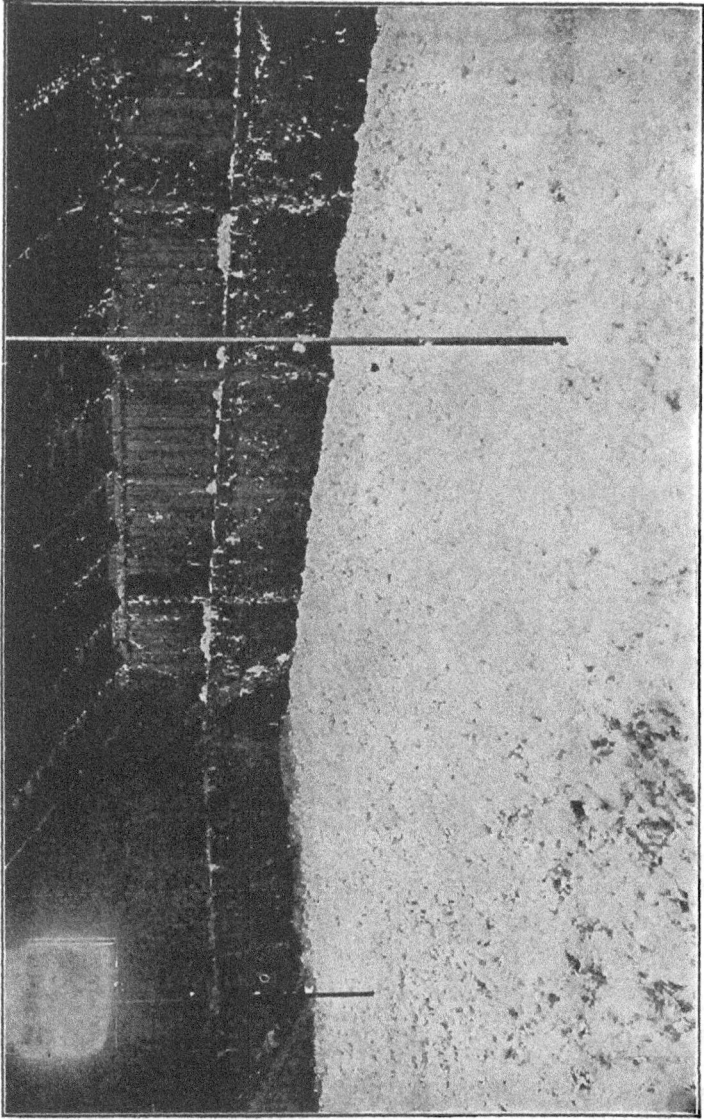

A valuable by-product of duck plants. The feathers from a duck will pay for the cost of picking. (Photograph from the Bureau of Animal Industry, U. S. Department of Agriculture.)

CHAPTER XIII

Establishing a Duck Farm

Much thought should be given to the location of a duck farm before deciding where to start; for its location with relation to markets, type of soil, shade and its proximity to water and the slope, or "lay of the land" may be the deciding factor in whether your duck farming will be a success or failure.

All of these will have direct bearing on health of the ducks, economy of marketing, labor and a general but very important bearing on the success of your enterprise.

Consult the County Agricultural Agent of the U. S. Agricultural Extension Service, who, in any section where ducks are raised, you will find well posted and a source of valuable information. He has had occasion to see the best results as well as the failures of men who make duck raising their business and he will be more than glad to pass this valuable information on to you.

The practically universal use of the Pekin as a meat duck on farms where market ducks are the object, would seem to indicate that this is the best choice for raising ducks for meat. If egg production is the object, the White Indian Runner is probably the best choice. The premium on white feathers, a valuable by-product of the duck farm would be enough to turn the scale in favor of a white breed if one were in doubt.

Ducks do best in a temperate climate. They do not mind cold although in extremely cold climates it will cost more for feed and for housing and brooding. But ducks do not do as well in extremely hot weather and so the temperate climate is most suitable.

A light sandy soil is best from a sanitary standpoint. This type of soil will purify itself more quickly than will heavier land, by rains and snow and so help to keep down disease taints and infection. A never failing water supply is very essential. If land can be secured through which runs a sizable stream with land on either side on which yards can be run down to the water, it is very desirable and will save much labor in the carting or piping of water and will also add to the health and vigor of the breeding ducks. If the land is sloping or rolling, so much the better, as this will help to drain and purify the runs without labor of ploughing or tilling.

The location with relation to markets is important. The nearer it is to a good market and a railroad station, the less time and expense will be consumed in transporting the dressed ducks or eggs from farm to market and grain, litter and other supplies to the farm. Not only the distance to market is important but the kind of road over which produce must be transported. If possible, locate on a state highway or one which is kept cleared of snow and which is kept always in good condition. While the local market may absorb most of your product there may be times when there will be an over supply and so it is wise not to locate your farm too far from one of the great cities where surpluses may be shipped without too much delay or length of time in shipment.

Do not depend on salt or even brackish streams for the ducks to swim in even if fresh water is supplied them; they will persist in getting some of the brackish water and while this may not actually injure them, it may hurt their appetites and lower the food consumption to the point of impaired growth; and quick

Dressed duckling. The main feathers of the tail and wings and the feathers of the neck part of the way from the head to the body are left on. The rest of the body is picked clean. (Photograph from the Bureau of Animal Industry, U. S. Department of Agriculture.)

growth through heavy and efficient consumption of feed is the most important factor in growing ducks for meat.

On the scale of business planned, depends largely the type of buildings required. In fact, if only a few ducks are to be raised, there will probably be buildings on the place such as barns, sheds or chicken houses that can, through a little remodelling be used very satisfactorily; or small colony type or shed roof chicken houses may be built. However, if the business is to be on a large scale, more complete and well planned buildings should be planned and constructed.

There must be buildings for laying and growing stock, incubator cellar, brooder houses, killing house, storage and mixing rooms for feed, supplies, etc. One story buildings are more suitable for ducks as they must have easy and convenient access to land. Before planning your buildings, visit other duck farms in the vicinity and consult your County Agricultural Agent.

That the sun may shine into the duck house for its sanitizing and drying effect, it is well to have it face the south although it is not as important as with hens from the standpoint of the ducks as they stay in the house very little except at night.

The floors in duck houses are usually of dirt but cement floors are better as they may be so much more easily cleaned and kept sanitary. Either way, the floor should be about a foot above the level of the ground and well covered with sand or litter. Shavings, sugar-cane litter or straw are materials generally used and there is little to choose between them. Litter in cold weather may be allowed to accumulate as the volume under foot will add to warmth but care must be taken to keep it reasonably dry and for this reason all the feeding and watering should be done on the outside of the house if that is possible. New litter is added as frequently as needed, depending on the weather and condition of the old litter.

When planning floor space in buildings, figure from three to four square feet per bird. The greater number together, the less feet of floor space for each bird. It is a question how many to run in one flock. Any number from 10 to 500 are run in flocks but usually from 100 to 150 is customary. If no more than this number are run together, there will be less cripples and fewer broken eggs but it is more expensive in building and labor to house them in the small units, where large numbers running into the thousands are kept.

Many duck growers use no nests at all but as a rule nests are provided on the floor level around the outside of the room against the walls. They should be about 12 inches wide in front, 15 to 18 inches deep with partitions a foot high. The partitions may be held in place by a 5″ by ⅞″ strip set on edge in front which forms the front of the nests, the wall of the building forming the back.

On this plant, the lay of the land was such that not all of the yards could be run down to the stream. So a shallow canal was dug from the stream through the yards which were without natural water frontage. (Photograph from the Bureau of Animal Industry, U. S. Department of Agriculture.)

CHAPTER XIV

General

Ample feeding space must be provided and trays 5′ long, 3′ wide and 4″ deep should be provided for each 50 ducks. There is expense involved in providing drains for the slop from watering devices for such time as the ducks must be watered in the house in cold weather, but it is money well spent. Locate each water container over a metal or wooden grill covering a tile or other drain for the purpose of conveying waste water, in the interest of keeping the litter as dry as possible. An iron pipe-line with a faucet at each water container will save much time and labor in watering a flock of ducks.

Yards for meat-type ducks should be approximately 50 by 300 feet for 200 ducks or about 75 square feet per bird. For ideal conditions one-third of this should be water and if possible, running water and fresh, not brackish. The land should be sloping if possible and sandy soil aids greatly in keeping the yards clean. The top surface should be sliced off with shovels and removed several times during the summer to still further insure against contamination.

Movable wire or slat fences about 2½′ high are all that is needed to confine the ducks as permanent fences make more work than when yards are cleaned and changed, from time to time, as may be desirable.

We are told by some that water for swimming is wholly unnecessary for growing or laying ducks but ducks being grown for breeding and the breeding ducks themselves will be benefited by a chance to swim every day. They develop better frames and muscles and fertility will average better when they have access to their native element.

Ducks grown solely for meat may not grow faster with access to water, but will have less mortality in the flock by developing stronger bodies and more disease resistance through better health.

The more active laying-type ducks require somewhat more yard room than the meat ducks although they will do well and produce profitably in extremely small yards. Above all things, ducks of any breed, size or age, require fresh, clean drinking water in plentiful supply.

Lights

It is well to equip any building in which young ducks are kept, with lights as they are readily scared and when frightened are very apt to stampede in a crazy rush and injure themselves by crowding and trampling. These lights need not be bright; a fifteen watt electric lamp being sufficient for a space 20 by 40.

Artificial lights also tend to increase egg production by giving longer opportunity for the ducks to eat, making possible more intake of nourishment. When lights have been used, they should not be discontinued especially during the laying season as changes of any kind tend to slow down laying for the flock.

Feeding fattening or yard ducks from the feeding track. (Photograph from the Bureau of Animal Industry, U. S. Department of Agriculture.)

CHAPTER XV

Exhibiting Ducks

Shows and expositions that feature departments for ducks can be of great value to duck breeders and fanciers. If one is interested solely in the growing of ducks for market, the publicity to be derived from displays of live specimens of your breed in spotless condition, attractively shown with decorations and information cards telling of the quality of your products, the sanitary conditions under which they are kept and raised and the trade name under which they are marketed, together with places where they may be purchased, will be sure to attract the attention of the public and should result in increased demand for your particular product.

If possible a display of the dressed ducks in a glass show case will add to the effectiveness or "eye appeal" of the exhibit. If refrigeration is not convenient, a little dry ice will keep the show-case cold for the duration of the show. Exhibiting where attendance of the general public is large is one of the cheapest and most effective methods of publicity for any food, and ducks are no exception to this rule. It will be worth looking into if there is a show or exhibition anywhere near your farm or the market which consumes your ducks. If you are engaged in the breeding of ducks for sale as breeders or for exhibition, the show or exhibition is equally, and perhaps more, important as a publicity method than to the commercial grower.

Premiums won in competition with other breeders tell the story of breed quality, graphically and effectively. If you issue a circular nothing will impress the potential buyer of your birds more than a printed list of prizes won in competition at shows and the bigger the show the more valuable the effect on the possible buyer.

If you are breeding ducks for fun and the satisfaction of improving your breed, there is no method of determining its advancement year by year as well as the measuring stick of the competitive show. By exhibiting in competition with other breeders, under good judges, you have an unbiased and authoritative official record of just how your birds compare in breed-type, color and markings with other breeders and it is invaluable to you in continuing your work of breed improvement; and this goes much farther in the minds of the public than any statements you might make, based on your own opinion. You may read all the books and study all the pictures, but nothing will fix in your mind just what is necessary in improving your breed as much as to send some to a show where there is apt to be competition in your own breed and then attend the exhibition in person, study your exhibit, compare the birds with those of your competitor, discuss it with the judge and then decide what your ducks lack and plan to remedy that lack. Most of all, meet the judge if possible and frankly discuss your birds with him. You'll find him well disposed and glad to talk over your problems. He'll tell you where your birds are strong and where weak. He has probably been selected for his knowledge gained through long experience and study of the breeds to which he has been assigned in the show and he'll be happy to explain why you lost if you did lose, and why your birds were placed over those of your competitors if you won. Talk with him in either case and you will be the gainer. The thrill of the blue ribbon is only secondary to the educational and advertising value of exhibiting whatever your object may be in breeding and growing ducks.

But use judgment in selecting the show at which you enter your birds. Find out if the management is sympathetic to the waterfowl department; there have been shows that did not welcome ducks as a part of the show and consequently were not cooperative and helpful either in the selection of the judge, care of the ducks while on exhibition and not willing to give them a fair deal in their location in the show room. Such shows should be avoided as it is expensive in both time and money to exhibit and the shows where you will be most benefited should be selected. Usually the premium list will indicate the attitude of the management in this respect and its perusal will help you to decide whether or not to show there.

If you plan on showing, look up a list of the big fairs and shows for the season; the dates are usually published in the poultry and farm papers. Then send to each one for a premium list. Look them over and then decide where you will reap the most benefit by exhibiting. Bear in mind the most important features from the exhibitor's standpoint, is competition and publicity. Find out in advance whether there is usually a good exhibit of ducks. Write the manager of the show for a list of last year's exhibitors and the breeds they showed. Most managers will be glad to give you this information. Then determine how many people usually attend the show. Total number of attendance is not all. If the show is held in a great city and the attendance is practically of city folks who are simply looking for something different from the usual city sights, the attendance from your standpoint is not as valuable as a smaller attendance but made up principally of country people who have become interested in the show through poultry and farm press advertising. This, you will readily see the reason for. Every person who lives in the country; whether on a farm or country estate or simply a residence in the country with from half an acre of land up, is a potential buyer of breeding birds and remember, the sale of breeding birds at breeding-bird prices is one of the prime advantages of exhibiting.

How important the show is in the eyes of the farm and poultry press is also a very important factor in its value to the exhibitor. When the show is over and all the value of the interest of the folks who attended the show is added up, if the awards are not published and the story of the show spread far and wide through the press, a lot of its potential value to the exhibitor is cut off. But if the winnings and the general story of the show are spread through the medium of the press they may reach 100 people to every one that saw your birds in the show; so it is easy to see that here is a vast supplementary benefit to those who make up the show through their exhibits.

As a rule the show in a big city is of greater publicity value than the one in a small town. But by the same token it costs more in entry fees and costs more to go there and stay through the show. If possible, go to the show and be in attendance for as many days as possible. This is not absolutely necessary as every show, worthwhile, has capable attendants who will receive your birds if shipped by express and will care for them well and return by express after the show is over.

When you decide when and at what show you wish to exhibit, send a post card to the show manager asking for a premium list. Send this request eight weeks in advance of the show dates, at least, as most shows close their entries a month or more in advance, in order that proper arrangements for cooping and manage-

Convenient feeding arrangements. At the right of the feeding track runs a water pipe with spigots and pans at frequent intervals. At the left are the feeding trays. (Photograph from the Bureau of Animal Industry, U. S. Department of Agriculture.)

ment may be made. With the premium list, you will receive entry blanks on which list your entries and return with check covering entry fees. The Show Manager or Secretary will, on receipt of your entry, send you shipping tags to be attached to your exhibit, in ample time for your shipment.

After the show is over, your work is not finished if you wish to capitalize fully on your exhibition. Let your local paper know when you ship your birds to the show and just what they are. When they return and if they have been fortunate and won prizes don't fail to favor your local paper with another news item. If you have no local paper look up the local correspondents of the nearest city papers and let them have the story. This is local news and they are always glad to have it. Most of the local correspondents are paid by the line and the more linage, the more pay. Someone interested in ducks is liable to read it and the result may be a direct sale and it is all added to your reputation as a breeder of ducks and this is valuable. Then, if you have won at the show, have a little circular printed and mail to as many duck breeders as possible.

Just a word in preparation of ducks for exhibition; and it requires only a word. All a duck needs in preparation for a show is full plumage, perfect health and spotless cleanliness. If your birds are not in full plumage at the time of the show, there is little you can do about it. Perhaps next year you can influence the moult a little, by the time you hatch the ducklings. Figure how old they were this year when out of moult at show time and if you wish to exhibit at the same show next year, hatch a month earlier—or later. It will not influence the time of moult as much as the difference in hatching time but it will make some difference. Have them in good flesh. Don't overdo it and have them so fat they cannot stand, as is sometimes done. But they should be fairly fat, especially in the meat breeds. The Runners should be thin and active. If there is opportunity for washing in clean water they will take care of that themselves. If not provide it; or thoroughly wash them in clean soap suds and as thoroughly rinse. Wash them at least twice, two days apart; and three times is better. But the best washing the ducks can have is in clean water and to let them do the job themselves.

In shipping to shows, have the coops light but strong. The sides of the coops should be solid with hinged, slatted tops and not too big. If shipping single ducks, have them not over ten inches wide and 20 to 28 inches long, depending on size of ducks, but have them high enough. They should be at least 20 inches high so that the duck may hold its head erect and be comfortable. Provide drinking cups if shipping more than four or five hours. If you are making your own coops, use a quart tomato can and fasten to the outside of the coop with an opening for the duck to put its head through to drink. If placed on the inside it is apt to soil the plumage and appearance. If two ducks are shipped together, the coop should be twice as wide and the same length and there should be two drinking receptacles. Have your shipping coops nicely painted with the name of your breed and your own name and address neatly and plainly lettered thereon. It all adds up to the sum total of the publicity which is the most important factor in all the advantages of exhibiting ducks.

CHAPTER XVI

Diseases of Ducks

Were the only readers of this book, breeders of ducks on a small scale; the fanciers, farmers and those who keep them for the production of meat and eggs for the family table, there would be little need of this chapter. For in small flocks of ducks kept under nearly normal conditions, disease is practically unknown.

However, as with other birds and animals when kept in large numbers as they must be for commercial purposes and under more or less artificial conditions, disease menace threatens in nearly the same ratio as management departs from normalcy and flocks increase in size. Ducks or geese are seldom if ever troubled with lice or mites. The following diseases, symptoms and methods of prevention and treatments by Dr. K. F. Hilbert of the Poultry Disease Laboratory at Farmingdale, Long Island, maintained by the New York State Veterinary College in cooperation with the New York State Institute of Agriculture, includes about all we know about diseases of ducks.

"Ducks raised under the usual commercial conditions are frequently subject to serious mortality. Breeder ducks are subject to most of the diseases of the young or market ducks and, in addition, to several other conditions that cause a mortality of a much less serious nature. Diseases causing the most serious losses are fowl cholera, anatipestifer infection, Salmonella infections (keel), and aspergillosis. Internal parasites except for an occasional tapeworm in old breeders are seldom found.

Fowl Cholera
(Duck Cholera, Hemorrhagic Septicemia)

"Fowl cholera is an infectious and contagious disease caused by *Pasteurella avicida*. Ducks, chickens, turkeys, pigeons, geese, cage birds, and wild birds may be affected. Many outbreaks follow adverse weather conditions, unsanitary conditions, overcrowding, and poor ventilation. The disease in ducks is usually acute but may be subacute or chronic.

Symptoms, Course, and Mortality

"The symptoms resemble those seen in chickens. They include increased thirst, loss of appetite, depression, and nervous twitching or jerking of the head. The temperature is elevated, and the neck and feet feel hot. A yellowish mucoid discharge which later becomes greenish and watery is noted from the bowels. The feathers around the vent usually become smeared with it. The eyes frequently are watery and often encrusted with dirt. The nostrils often contain a slimy to gelatinous exudate. The muscles are congested and the skin appears pinkish. Swollen feet or joints are seen in most outbreaks.

"The course of the disease usually is acute; losses up to 90 per cent may occur in a week or two. Those running a chronic course usually have localizations in the joints, causing a lameness, or in the airsacs, which result in emaciation and respiratory symptoms. Birds affected with the chronic form of the disease may live for several weeks. Mortality in ducks under four months of age may be more than 90 per cent, while the mortality in breeder ducks seldom reaches more than 50 per cent.

Autopsy Findings

"The autopsy findings in ducks are similar to those in chickens, although usually they are more pronounced. Often the heart is enlarged and studded with petechiae along the coronary fat. The pericardium may be thickened and the cavity may contain a yellowish fluid in which flakes are suspended. Petechiae may be found in the liver, on the lining of the body cavity, and in the gizzard fat. The liver is slightly swollen and sometimes contains minute whitish areas which give it a mottled appearance.

"The nasal sinuses are congested and often contain a gelatinous exudate. The lungs may be congested; occasionally consolidation is present. The airsacs are usually thickened slightly and have an abnormal yellowish opacity. The intestines are slightly or severely hemorrhagic, and the contents are semi-liquid and mucoid in consistency. The feces are yellowish or greenish. The blood vessels of the intestine are usually engorged as are those in the mesentery. The spleen and kidneys are usually swollen. The muscles are dark in color.

"In the chronic form of the disease, affected joints may contain a reddish serous, whitish creamy, or yellowish caseous exudate. The airsacs are usually thickened and contain large amount of yellowish caseous exudate. Birds affected with this form of the disease are always emaciated.

Diagnosis

"Although history, symptoms, and lesions may indicate the disease, bacteriological examination is necessary for positive diagnosis. Fowl cholera must be differentiated from such acute infections as anatipestifer infection and from poisoning. The color of the feces cannot be considered as an aid to diagnosis.

Prevention and Treatment

"Chemically killed autogenous bacterin 1, 2, 3, 4 is a means of preventing this disease even when the birds are kept under the insanitary conditions which prevail on the ordinary duck farm. The bacterin is injected subcutaneously, 1 cubic centimeter of a 48-hour growth, whole-culture bacterin is used for ducklings up to seven weeks of age, and 2 cubic centimeters for older ducks.

"Skidmore's (1932) observation on the common house fly as a possible carrier emphasizes the need of prompt removal of sick birds and disposal of dead birds by burning, if possible, instead of by burial: otherwise the diseased carcasses may be dug up by dogs and other animals and serve as a reservoir of infection. This is important because of the very large number of flies on duck farms. Strict sanitation should be employed.

"The only treatment that appears to have any value is large doses (from 15 to 20 cubic centimeters) of antihemorrhagic septicemia serum. This treatment sometimes appears to save about half of the affected ducks. The cost of this treatment, in addition to the extra length of time required to obtain marketable weight of affected birds, makes destruction of the sick birds more economical.

Anatipestifer Infection

"Anatipestifer infection is a highly infectious disease due to *Pfiefferella anatipestifer*. The disease is acute, death often occurring in from six to twelve hours after symptoms are first noticed. Some birds may develop a more chronic disease which terminates with death after several days. Ducks under two weeks of age and breeder ducks appear to be refractory to this disease. The highest mortality has been in ducks from five to ten weeks old. The disease was first described by Hendrickson and Hilbert (1932) in 1932. A disease which appears to be identical was described by Graham, Brandley, and Dunlap (1938) in 1938.

Symptoms

"The first symptoms of the disease are depression, manifested by a sleepy attitude, and ruffling of the feathers. There is a severe diarrhea, the discharges being of a distinct greenish color. The affected birds are often unable to stand and there is continual bobbing and jerking of the head. There is often a serous discharge from the eyes. Later, complete prostration occurs, the birds lying on the ground or floor of the pen unable to raise their heads.

Autopsy Findings

"The lesions are those of a rapidly developing septicemia, Petechial hemorrhages are commonly found on the serous surfaces of the liver, heart, and body wall. In nearly all cases, the prominent lesion is a yellowish gelatinous exudate that covers the liver and adheres so closely to it that it can be removed only with difficulty. A similar exudate causes adhesions of the pericardium to the heart wall. The coronary fat and tip of the heart are sometimes oedematous. The spleen is usually as least slightly enlarged and presents a peculiar brown and white mottling. The kidneys are congested. Most birds exhibit a marked hemorrhagic enteritis. Large quantities of blood-stained exudate are found on the intestinal mucosa. The lungs are highly congested and occasionally pneumonic. The blood is dark in color and often uncoagulated.

Diagnosis

"The yellowish fibrinous exudate covering the liver and heart is suggestive of this disease, but a definite diagnosis must rest on the isolation and identification of the causative organism from the blood and internal organs of dead ducks.

Treatment and Control

"Treatment is of no avail. Experimental work indicates that a chemically killed bacterin may be an aid in control. Production of such a bacterin appears to be impractical at present as the organisms isolated from Long Island ducks grows so poorly on artificial media that the difficulty of making the bacterin makes its cost practically prohibitive. Rigid sanitation and destruction of all affected ducks as soon as they are noticed seems to offer the best means of control at the present time.

Salmonella Infections
(Keel)

"Salmonella infection is a septicemic disease of young ducks caused by *Salmonella anatis, Salmonella typhimurium,* or *Salmonella enteritidis* Gaertner. The mortality may reach 50 per cent during the first three weeks of life. The average mortality on Long Island does not exceed 30 per cent. High mortality occurs under unsanitary conditions and is very often complicated with rickets. During the past two or three years, the disease has become relatively infrequent on Long Island.

Symptoms

"Ducklings die suddenly, hence the name *keel*. Others have watery eyes; sometimes the eyelids are pasted together. The nostrils are wet or contain mucus. Many ducklings exhibit difficulty in breathing. Some gasp for air. Rapid breathing is a common symptom. Coughing, sneezing, or mucus 'clicks' are frequently encountered. Death may occur before symptoms are seen, while sometimes life lingers for two or three days. Ducklings in an infected lot are inactive and somewhat depressed.

Autopsy Findings

"Frequently, the only gross lesion seen is an enlargement of the liver which may appear yellowish. Congestion of the lungs and sometimes pneumonia are seen. Many birds have a serous or thick mucous exudate in the nostrils or nasal cavities. Extensive catarrhal enteritis is usually found. A few show increased pericardial fluid and a flabby heart wall.

Diagnosis

"Diagnosis may be made by isolating the causative organism. It is necessary to differentiate this disease from respiratory infections, other bacterial diseases, and poisoning.

Prevention and Treatment

"Strict sanitation is required both in the brooder and the incubator. A well-balanced ration containing adequate amounts of cod-liver oil should be fed to both breeders and ducklings. Control of the disease in breeding flocks through eliminating reactors to the agglutination test has been unsatisfactory, apparently because the spread through this means is not the important factor in the dissemination of the disease. Cultural examination of eggs from reactors (carriers) gave negative results and all young ducklings hatched from this group of ducks failed to contract the disease. J. S. Garside and R. F. Gourdon (1940) found that breeding experiments carried out at the laboratory and the cultural examination of eggs laid by carrier birds gave negative results. This, together with observations on the course of the disease in the field, tends to indicate that egg transmission was not the most important factor in the dissemination of the infection, and that adult carriers served only as a means of introducing the disease occasionally, in spite of the fact that on post-mortem examination of 68 carriers the organism was successfully isolated from the ovary on 37 occasions. Possibly such birds may lay comparatively few eggs with a low percentage of infection or none at all. Vermin appear to play an important part as natural reservoirs and disseminators of infection.

"Feeding pellets in hoppers rather than a wet mash on open flats appears to aid in the control of this disease because it keeps the infected droppings out of the food. The method of supplying water should be such that it cannot be contaminated by the ducklings and their droppings. Treatment of affected ducklings is unsatisfactory. Sick and dead ducklings should be disposed of as soon as noticed.

Brooder Pneumonia
(Aspergillosis)

"Brooder pneumonia is a disease of ducklings caused by the mold *Aspergillus fumigatus*. It is often introduced through contaminated food and litter. It may appear to spread rapidly in a closely confined brooder house. Respiratory symptoms are practically the only ones seen.

"The disease affects the lungs, airsacs, upper air passages and sometimes the mouth. The lungs often contain caseous areas and the airsacks frequently are lined with greenish areas. In some birds, the airsacs may contain many small whitish bodies, tightly adherent to the airsac wall.

"The mold can easily be demonstrated in smears from the greenish areas. In denser lesions, it is difficult to demonstrate the mold except by cultural methods on wort or acid agar.

"Prevention lies in keeping the mold from the ducklings. Moldy food and litter must be strictly guarded against. Wet spots in the pen should be eliminated. Wet areas around the watering containers should be prevented as much as possible by removing the litter around these areas every few days and replacing it with new clean litter. Treatment is unsatisfactory.

Botulism
(Limber-neck)

"Botulism is a poisoning produced by eating decomposed animal or vegetable matter or maggots that have developed on such matter. This poison is produced by the germs of botulism which are widely distributed in soil. Unlike most disease-producing germs, this one cannot multiply in the body. It multiplies in decomposing carcasses and in moist food stuffs in which it produces a powerful poison. In some of the Western States, losses in wild ducks have been heavy as a result of feeding upon decomposing vegetable material in swamps and on the bottom of shallow stagnant pools.

"The symptoms of botulism appear in from eight to twenty-four hours after the poisonous food is eaten. If a small amount of poison is eaten, the disease is mild and affected birds usually recover. If large amounts are eaten, it is fatal. The poison may be spread from droppings of affected birds and carcasses of birds dead from the disease.

"Post-mortem examination reveals no visible changes in the internal organs.

Symptoms

"The mild form of the disease causes leg weakness. Affected birds usually recover in a few days.

"The more severe form of botulism produces sleepiness. Paralysis of the neck muscles appears early and, as a result, the birds are unable to hold their heads erect. The wings and legs may become completely paralyzed. Affected birds often lie outstretched on their sides. They often remain in this condition for many hours before they die. The feathers over the back may often be easily plucked out.

Treatment and Control

"Any contaminated food should be removed. The carcasses of dead birds should be burned and the droppings of affected birds removed from the pens and yards. Epsom salts at the rate of 1 pound to each 500 pounds of birds, given in the drinking water, is recommended. Half-ounce doses of castor oil may be given. Valuable birds may be given antitoxin, types A and C as a preventive. The high cost of antitoxin, and the fact that it does not help much after symptoms have appeared, limits its use to valuable birds in the early stages of the disease.

Paralysis of the Penis

"Paralysis of the penis is seen rather frequently in drakes. It occurs most frequently in breeding flocks. The specific cause is unknown. The addition of extra drakes will prevent the occurrence of new cases in some breeding flocks but not in others.

Symptoms

"The penis protrudes from the vent and cannot be retracted. In a few days it becomes dry, dirty, scabby, and will, in most instances, finally slough off completely. The drake will continue to lose weight and may die. Those that live remain in poor condition for months. In the winter months, it is not unusual to have the penis freeze to the ice or other objects when the bird sits down after having been in the water.

Autopsy Findings

"In early cases, the birds will be in good flesh. Prolonged cases result in emaciation. The penis will show varying stages of necrosis and infection and will be protruded from the vent. Bacteriological examinations and direct inoculations fail to reproduce the disease.

Prevention and Treatment

"Treatment is of no value. Since overuse may be a factor in the causation of this disease, it is well to provide at least one drake for each five or six ducks in flocks when the trouble occurs.

Impacted Oviduct

"Impacted oviduct is due to an inflammation which causes the glands to produce abnormal amounts of albumin. The material changes its appearance and collects as a cheesy mass. Infections, foreign matter, and impacted eggs may cause the stimulation of the glands. Flukes in the oviduct have not been found in Long Island ducks.

Internal Layers

"Fully formed eggs are found in the body cavity. One theory is that after the egg is fully formed the aviduct muscles contract and carry the egg forward into the body cavity rather than backward through the vent. Such birds should be eliminated since removal of such eggs through the body cavity provides only temporary relief. Many such birds are found only at autopsy.

Eversion of the Oviduct
(Prolapsus, Blow Out)

"Eversions often follow the condition of 'egg-bound.' The oviduct is inflamed and together with the cloaca is pushed out through the vent. If discovered in time, the condition can frequently be reduced. The parts should be well washed, then gently pushed back into place with clean fingers."

—*Duck Growing Bulletin 345.* Cornell University 1936.

CHAPTER XVII

Sexing Day-Old Ducklings

Recently much interest has developed in the matter of sexing day-old chicks and to some extent with day-old ducklings.

While the early determination of sex in ducks is much easier and very accurate, its practical value is limited. Ducklings of the meat breeds are usually hatched for market purposes and no differentiation made in sex. Hatcherymen specializing in day-old ducklings of the laying breeds may find it practical to offer only female ducklings to customers who desire to start flocks for egg production and who may not care to bother growing and marketing drakes.

By eliminating the drakes immediately the customer is spared the transportation costs, extra brooder space and bother of sexing later on. Cost of sexing, premium charged for 100% females and disposition to be made of the drakes, will usually determine each individual case.

The method of procedure as given is taken from an article on "How to Sex Day-Old Ducklings," by Violet K. Tallent, *1934-35 Year Book* of The British Duck Keepers Association. Credit is due the author of the article for directions on sexing and to Mr. Ruff Jackson for accompanying illustrations.

"To avoid chilling the newly-hatched ducklings always sex them in a warm room.

"Only drakelets are discarded. Any mistake will mean the inclusion of a drakelet among the ducklets.

"The correct handling can be practiced on dead

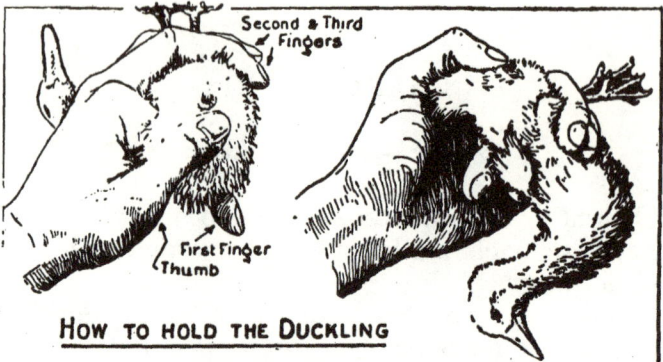

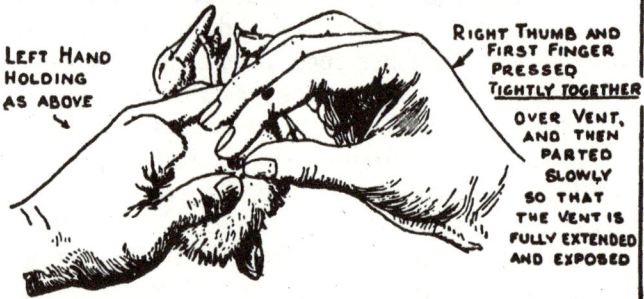

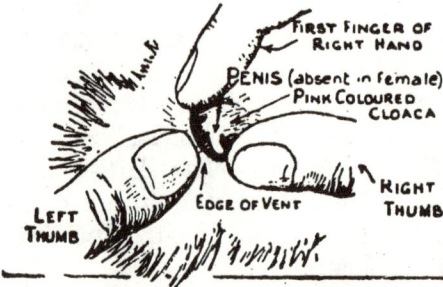

Chart for Sexing Ducklings and Goslings

ducklings but, to prevent the spread of disease, disinfect the hands before passing on to live ducklings.

"The difference between the sexes is easily seen, but some people find a watchmaker's magnifying glass helpful.

"If the beginner has difficulty in sexing a duckling he should put it aside and return to it later after sexing a few more. Ducklings can be hurt by too much handling.

"At hatching the cloaca is often full of greenish excreta. Some people find it more pleasant to sex the ducklings at a few days old.

"1. Take hold of the duckling's tail between the first finger and thumb of the left hand so that its breast is upwards and away from you and its head hangs down (*Fig. 1*).

"2. Then stretch the cloaca longitudinally by bending the duckling gently but firmly over the first finger and holding its body there with the middle, third and little fingers.

"3. Now stretch the cloaca transversely and so force out the copulatory organ, if it is there, by closing the thumb and first finger of the left hand, placing them while closed on the cloaca and then pasting them slowly. It is important that the thumb and first finger are kept together until they are placed on the cloaca because before they are parted, they must catch the edge of the cloaca and exert sufficient tension to force out the copulatory organ. This organ looks like a pinkish root tip and its presence denotes a drakelet. No such organ is seen in the cloaca of the ducklet (Figs. 2 and 3).

"Geese can be sexed in the same way."

—Seventh edition by J. M. Hunter and J. C. Scholes, *Profitable Duck Management*, Beacon Milling Company, 1946.

CHAPTER XVIII

Ten Favorite Recipes for Preparing Duck

"Three Big Meals." Courtesy Franklane L. Sewell.

ROAST DUCK

A 5 lb. duck	1 cup seedless raisins
Salt, pepper, 1 garlic clove	1 cup orange juice
3 cups apples, quartered and pared	

Wash and clean duck, season and rub all over with clove of garlic, fill with apples and raisins. Place in roasting pan and cook uncovered in slow oven (325° F), allowing 20 to 30 minutes per pound. Baste every 15 minutes with orange juice. Serve with cranberry or currant jelly. Serves five.

DUCK WITH SAUERKRAUT

A 6 lb. duck	2 qts. sauerkraut
Salt, pepper, 1 garlic clove	1 cup water
3 tablespoons sugar	

Prepare duck as for roasting. Place in pan adding sauerkraut, water and sugar. Cover and bake in moderate oven (325° F.) until tender. Allow 20 to 25 minutes per pound. Serves six.

CANETON A LA BRIGARDE

A 2 lb. duckling	Juice of 1 orange
⅔ cup carrots sliced thin	¾ cup currant jelly
½ cup onions sliced thin	¼ cup kümmel
Rind of ¼ orange	2 pieces toast

Roast duckling in moderate oven (325°) until tender. Cut orange rind into thin strips and cook in water until slightly tender. Fry vegetables in fat from duckling, remove from pan and add orange juice, jelly and kümmel to remaining fat in pan. Mix. Place cooked duckling on toast, surround with orange rind, onions and carrots. Serves two.

FRICASSEE OF DUCK

A 5 lb. duck	1 bay leaf
Salt, pepper, flour	2 tablespoons chopped
6 tablespoons fat	onions
1 cup chopped mushrooms	

Clean and cut duck into serving portions. Season and dredge with salt, pepper and flour. Brown duck with onion and bay leaf in 4 tablespoons of fat. Cover with water and simmer in covered pan until tender. Saute mushrooms in remaining fat. Add to duck and cook 10 minutes longer. Serves five.

DUCKLING AND FRUIT SALAD

2 cups orange sections	2 cups cooked duckling diced
½ cup French dressing	1 cup grapefruit sections

Mix all ingredients together. Serve chilled on lettuce. Serves six.

BRAISED DUCK

A 4 lb. duck	1 teaspoon salt
3 slices bacon	½ teaspoon pepper
1 diced carrot	4 cups boiling water
1 onion stuck with whole	1 small turnip, diced
cloves	2 tablespoons melted fat
½ teaspoon thyme	4 tablespoons flour
2 tablespoons minced parsley	¼ cup cold water

Prepare duck for roasting. Place in kettle with carrot, onion, parsley, thyme, salt, pepper and bacon. Cover with water and simmer until duck is tender. Remove duck from liquid. Saute turnip in fat until brown. Drain and add turnip to stock. Cook until tender. Strain stock. Mix flour and cold water, add slowly to stock, stirring constantly. Pour the gravy over duck. Serves six.

DUCK A LA CREOLE

2 tablespoons fat	½ cup finely chopped celery
1 tablespoon flour	2 tablespoons chopped sweet
2 tablespoons finely chopped	pepper
ham	1 tablespoon minced parsley
¾ teaspoon salt	1½ cups consomme
¼ teaspoon pepper	1 whole clove
2 tablespoons minced onion	¼ teaspoon mace
2 cups diced cooked duck	

Melt fat, blend in flour and stir in ham. Season with salt, pepper, celery, onion, pepper and parsley. Cook for 2 minutes, add consomme, mace and clove. Simmer about an hour. Strain sauce and stir in duck. Reheat and serve with fried hominy. Serves four.

PINEAPPLE DUCK

A 3 lb. duck	1 No. 2 can sliced pineapple
2 cups boiling water	(diced)
2 tablespoons soy sauce	Salt and pepper

Cut duck into serving pieces, cover with boiling water and simmer about 1 hour (until nearly tender). Add salt, pepper, soy sauce, can of pineapple (including syrup) and cook 30 minutes longer. Serves four.

CHESTNUT DUCK

A 3 lb. duck	Salt and pepper
2 cups boiling water	1 tablespoons soy sauce
6 cups chestnuts	2 cups sliced mushrooms

Cut duck into serving pieces. Cover with boiling water and simmer 1 hour. Put chestnuts in cold water and discard those that float. Simmer chestnuts for 5 minutes then drain and peal. Add chestnuts, mushrooms, salt, pepper and say sauce to duck. Cook slowly about 30 minutes longer. Serves five.

DUCK SOUP

A 5 lb. duck	4 dried orange peels, ground
1 teaspoon dried onion flakes	3 stalks celery cut fine
About 3 quarts boiling water	1 small onion
ter	4 tablespoons soy sauce
•	Salt and pepper

Cover duck with water. Add orange peel, dried onion, celery and onion. Simmer for about three hours. Remove duck and add remaining ingredients. Serves ten. Duck may be fried later.

INDEX

PART ONE—GEESE

PART TWO—DUCKS

Printed in the USA
CPSIA information can be obtained
at www.ICGtesting.com
LVHW041615240923
759037LV00002B/94